Mineral Resources Economics 2

SCIENCES

Energy, Field Directors – Alain Dollet, Pascal Brault

Raw Materials and Materials for Energy,
Subject Heads – Olivier Vidal and Frédéric Schuster

Mineral Resources Economics 2

Issues and Action Levers

Coordinated by
Florian Fizaine
Xavier Galiègue

iSTE

WILEY

First published 2021 in Great Britain and the United States by ISTE Ltd and John Wiley & Sons, Inc.

ISTE Ltd
27-37 St George's Road
London SW19 4EU
UK

www.iste.co.uk

John Wiley & Sons, Inc.
111 River Street
Hoboken, NJ 07030
USA

www.wiley.com

Library of Congress Control Number: 2021940276

British Library Cataloguing-in-Publication Data
A CIP record for this book is available from the British Library
ISBN 978-1-78945-025-5

ERC code:
SH1 Individuals, Markets and Organisations
SH1_12 Agricultural economics; energy economics; environmental economics
PE10 Earth System Science
PE10_10 Mineralogy, petrology, igneous petrology, metamorphic petrology

MIX
Paper from responsible sources
FSC www.fsc.org FSC® C013604

Contents

Introduction xi
Florian FIZAINE and Xavier GALIÈGUE

Part 1. Stakes 1

Chapter 1. Toward a New Geopolitics of Raw Materials in the Energy Transition 3
Emmanuel HACHE, Gondia SOKHNA SECK, Charlène BARNET, Samuel CARCANAGUE and Fernanda GUEDES

1.1. Introduction 3
1.2. Measuring the criticality of raw materials and geopolitical risk 5
1.2.1. Criticality, strategic materials and risks 5
1.2.2. The absence of a homogeneous theoretical framework 6
1.2.3. Criticality matrices 7
1.3. The geopolitics and geo-economics of raw materials in the energy transition 11
1.3.1. From measuring pressures on reserves to taking geopolitics into account in measuring criticality 12
1.3.2. Fear of cartelization or monopoly in commodity markets 13

1.4. How can we manage strategic materials supply risk? 24
1.4.1. The role of public policies 25
1.4.2. The issue of strategic stocks 27
1.4.3. Foreign investment through national companies 28
1.4.4. The logic of the Chinese barter 30
1.5. Conclusion: toward a new resource nationalism? 30
1.6. References 32

Chapter 2. Legal Issues Regarding the Sustainable Management of Territorial and Extraterritorial Mineral Resources 39
Stephanie REICHE-DE VIGAN

2.1. National law regarding territorial mineral resources: the decisive issue of ownership 42
2.1.1. Ownership over mineral resources at the core of mineral law 42
2.1.2. A form of mineral ownership that may limit the government's capacity to regulate the extractive sector for environmental reasons 47
2.2. International law regarding territorial mineral resources: the central role of state sovereignty 52
2.2.1. The principle of permanent sovereignty over natural resources for the benefit of international trade 52
2.2.2. A principle challenged by indigenous peoples' rights to lands, territories and resources 55
2.3. International law regarding extraterritorial mineral resources: exploitation "for the benefit of mankind as a whole" . . . 56
2.3.1. The legal status of the seabed and the subsoil, determined by states' interests in the exploitation of mineral resources 57
2.3.2. The legal framework for the exploitation of Antarctic mineral resources, determined by ecological considerations . . . 63
2.4. For a sustainable management of mineral resources 65
2.5. References 68

Chapter 3. Mining and Societies 71
Michel DESHAIES

3.1. Introduction 71

3.2. Mines as a factor of settlement and landscape transformation 72
3.2.1. Mining and the population 72
3.2.2. Mines, landscapes and the environment in pre-industrial times 74
3.3. Mining in the Industrial Age 76
3.3.1. The transformations of the industrial energy system 76
3.3.2. Birth and development of coalfields 78
3.3.3. Conquest and development of new metal deposits 79
3.4. Contemporary mining transformations and challenges 81
3.4.1. Geographic trends in mining 81
3.4.2. Decline and changes in former mining regions 82
3.4.3. Extraction boom and risks in new mining regions 85
3.4.4. The limits of "responsible" mining 92
3.5. Conclusion 95
3.6. References 96

Part 2. Action Levers 101

Chapter 4. Maintaining or Even Developing the Mining of Mineral Resources in Europe: The Case of Wallonia (Belgium) 103
Johan YANS

4.1. Introduction 103
4.2. Geological resources in Wallonia 104
4.2.1. Extraction of mineral materials other than metals 104
4.2.2. Metal extraction: a problem on several spatiotemporal scales 105
4.3. Extension of sites/quantity of mining? 106
4.3.1. Exploit existing and well-characterized metal resources/reserves 106
4.3.2. Promoting a short circuit 107
4.3.3. Promoting alternatives to the sometimes deplorable extraction conditions in some regions of the world 108
4.3.4. Stimulating the local economy/employment 108
4.3.5. (Re)discovering a degree of supply independence for the industry 108
4.3.6. Creating the "substitution threat": knowing that local potential exists 109

4.4. Decrease in sites/quantity of operations 109
4.4.1. Lack of local skills (being addressed) 109
4.4.2. NIMBY syndrome 110
4.5. Some levers for action 115
4.5.1. Responsible extraction. 115
4.5.2. Popularizing 115
4.5.3. Strengthening the administration and defining a clear public strategy 116
4.5.4. Consulting 117
4.5.5. Collaborating (private–public). 118
4.6. Conclusion 118
4.7. References. 119

Chapter 5. Substitution: Promises, Principles and Main Constraints 121
Florian FIZAINE

5.1. Introduction. 121
5.2. Main economic foundations of substitution 122
5.2.1. The demand curve 123
5.2.2. The horizons of substitution: short, medium and long term 124
5.2.3. The shortcomings of the classical demand curve 125
5.3. Elements, components, systems: what are we really substituting? 125
5.3.1. Altenpohl hierarchy and principal forms of technical substitution 126
5.3.2. Normative substitution: what to substitute for? 127
5.4. The main obstacles to substitution. 129
5.4.1. Technical obstacles 129
5.4.2. Economic obstacles. 130
5.4.3. Barriers related to the physical availability of the resource. 131
5.4.4. Cultural and historical barriers. 132
5.4.5. Regulatory barriers 133
5.5. Other aspects to be taken into account 134
5.5.1. Impact of competition and industrial strategies 134
5.5.2. Is economic substitution also an ecological substitution? 135
5.6. References. 136

Chapter 6. Resource Consumption and Decoupling 139
Thierry LEFÈVRE

6.1. Introduction 139
6.2. Global use of resources 142
6.3. Material consumption indicators 145
6.4. Decoupling the economy from resource consumption 149
6.4.1. Evidence of decoupling 149
6.4.2. Saturation of resource use 152
6.5. Responsibility for resource consumption 154
6.6. Conclusion 156
6.7. References 159

Chapter 7. The Economics of Recycling: Ambitions, Myths and Constraints 163
Alain GELDRON

7.1. The recycling economy, an ancient history 163
7.2. Geological and urban mines, similarities and differences in logic 165
7.3. Understand the definitions and indicators of recycling in order to express its performance 167
7.4. A limited deposit because we can only recycle what we have consumed 170
7.5. Multiple factors influencing recycling and its effectiveness . . . 173
7.6. The technical constraints of metal recycling 176
7.6.1. Preparation of materials 177
7.6.2. Recycling of base metals 178
7.6.3. Recycling of specialty metals 179
7.7. Environmental benefits of recycling 181
7.8. Conclusion 182
7.9. References 183

Chapter 8. Low-tech: A Path Toward the Necessary Metallic Sobriety? 187
Philippe BIHOUIX

8.1. Cornucopians versus doomsdayers 187
8.2. The circular economy, mission impossible? 190
8.2.1. Invisible dematerialization 191
8.2.2. The systemic issue between energy and resources 193

8.2.3. The constraints of recycling 194
8.3. Toward a metallic frugality 195
8.3.1. Sobriety above all. 196
8.3.2. "Advanced" eco-design 198
8.3.3. Moderate mechanization 199
8.4. A possible and desirable transition 200
8.4.1. The role of the public authority, at all scales 200
8.4.2. Finding the right scale 202
8.4.3. Humans, the key to "repairability" and optimal recycling 203
8.4.4. Positive impacts. 204
8.4.5. A "happy" transition or nothing 205
8.5. References. 206

Conclusion 209
Florian FIZAINE and Xavier GALIÈGUE

List of Authors 221

Index 223

Introduction

Florian FIZAINE[1] and Xavier GALIÈGUE[2]
[1] *IREGE, Savoie Mont Blanc University, Annecy, France*
[2] *LEO, University of Orléans, France*

We have seen in Volume 1 of this book that the context of mineral resources involves strong interactions between the material and physical constraints of extractive industries and those of the economy and social representations. These interactions are now strong enough for us to consider that we have already entered a new geological era, the Anthropocene era, in which the metabolism of the planet is shaped by human activity. However, unlike other geological eras, if the Anthropocene generates evolutions that are unintentional consequences of human activities, these could be voluntarily influenced – on the condition that public opinion becomes aware of this and assumes the consequences of voluntarist policies in this field.

A large part of this volume, Volume 2, will be devoted to examining the leverage available to us to make the energy transition compatible with the use of our mineral resources. However, before examining them, it will also be necessary to address a number of issues related to the legal regimes governing access to mineral resources, the geopolitical interactions between producing and consuming countries and the impact of extractive activities on populations.

Mineral Resources Economics 2,
coordinated by Florian FIZAINE and Xavier GALIÈGUE. © ISTE Ltd 2021.

I.1. The stakes: between legal rules, interdependence of nations and interests of local populations

First of all, it goes without saying that the perpetuation of a growth model based on the continuous growth of raw material needs, to which will most probably be added those (perhaps one-off) of the energy transition, places resource-importing countries in a difficult situation. In this context, the multiplication of criticality studies of raw materials testifies to the growing concern of countries and companies that rely heavily on international flows of natural resources. The contribution of *Emmanuel Hache and his co-authors* (Chapter 1) clearly shows that these criticality studies nonetheless bias reality by displacing consumers at the mercy of all-powerful producer countries. Yet this interdependence is often mutual, and it is therefore necessary to weigh the supply risk by also analyzing the diversity of exports from producer countries. Moreover, in this giant chess game, the availability of national resources (even untouched ones), the threat of the substitution of an alternative resource or a supply route from a cooperative country can certainly reduce the risk of supply disruption even in the case of a high level of dependence. Nevertheless, the reduction of the need for resources (sobriety and material efficiency) always seems less uncertain than securing access to primary resources.

Moreover, access to mineral resources is fundamentally shaped by legal rules and regimes, as demonstrated by the contribution of *Stéphanie Reich-de Vigan* (Chapter 2). Important differences remain between the legal regimes in the area of territorial and extraterritorial mineral resource ownership, which are not neutral with regard to incentives to extract or protect the environment. If our socio-cultural models focus on the economic part of sustainable development, it is likely that legal regimes for resource extraction will be brought into coherence to reflect this overarching objective. This will probably be to the detriment of the environment and by accentuating anthropic pressure on the last natural areas that have been relatively spared until now (Antarctica, the seabed in international waters).

There are, however, other options highlighted by the example of New Zealand. Indeed, New Zealand is one of the emblematic countries that have exploited the extension of the border, i.e. developed by relying on an adjacent geographic region exceptionally endowed with natural resources and characterized by a low human:earth ratio balance (Barbier 2011). However, when the United States continues to legally open other borders across the board – such as the treaty legalizing the private exploitation of asteroids – New Zealand opted for a different choice in 2017. Indeed, under pressure from the Iwi (Maori), who give the Whanganui River a different cultural representation from the British, the latter has been given a legal personality, allowing it to defend its interests in legal proceedings with the help of two lawyers. One can legitimately think that this will undoubtedly modify the future extractive trajectories near the river. This is again a good example of the importance of feedback from the human sciences on the opposing pressure by humans on the biosphere and the geosphere.

Within this framework, the study of the occupation of space, settlements and territories occupied by mining activity can provide adequate information on the parameters likely to modify these trajectories. This is what *Michel Deshaies* is trying to do in his contribution (Chapter 3). By these means, he confirms the absence of determinism on the trajectory of this anthropogenic pressure. He also notes the growing difficulties in perpetuating the current extraction model in the face of local populations increasingly hostile to mining projects. Here again, we note that conflicts related to mining projects are essentially present in North, South and Central America, even though a significant share of extraction is also located in Africa and other OECD countries. This is evidence that social organization, institutions and culture play a significant role in maintaining or inhibiting the widespread extraction of mineral resources.

I.2. The leverage for action: what can be expected and what remains to be explored

What leverage do we have at our disposal to achieve a more sober economy and sustainable development? One of the lessons that can be drawn from the previous chapters is that the existence of strong interactions between the various constraints affecting mineral resources prevents us from considering that we could free ourselves from them by resorting to univocal, purely technical, economic or societal solutions: proposing a simple solution to a complex problem is doomed to fail. In this volume, we examine what we can potentially expect from the deployment of several forms of leverage often evoked in the literature: domestic extraction, substitution, recycling, material efficiency (or decoupling) and low-tech solutions combined with sobriety.

To begin with, the maintenance or development of national or European domestic extraction is often mentioned both by certain geopolitical experts and by strategic committees, such as COMES (*Comité pour les métaux stratégiques*, Committee for Strategic Metals), concerned with reducing the impact of supply disruptions. This same recommendation is also sometimes made, this time in order to preserve the environment. However, although the stated objective is environmental, this leverage is rarely formulated by environmentalists themselves. We can think in particular of Guillaume Pitron's latest work for the general public (2018) where the author calls for higher national production to avoid a delocalization of the ecological impacts of the energy and digital transition (in his chapter "The end of the last sanctuaries"). This leverage would be both an effective way to relocate negative mining externalities while reducing them significantly (thanks to more respectful local extraction conditions), and a great support to raise consumers' awareness of the environmental cost of their lifestyle. The author even views it as a good way to increase consumer pressure on emerging countries to change their socially and environmentally disastrous extraction techniques. He criticizes environmental NGOs for not understanding the true impact of the energy transition:

> Environmental NGOs are showing a certain inconsistency as they denounce the more sustainable effects of the new world that they themselves have called for. They do not admit that the energy and digital transition is also a transition from oil fields to rare metal deposits, and that the fight against global warming calls for a mining response that must be assumed. (Pitron 2018)

While the latter argument may appear to be valid in the case of an energy transition in the form of "green growth", the argument for a national mining revival for the reasons cited above seems to suffer from at least four pitfalls:

– First, the revival of domestic mining supply suffers from multiple problems of perception by the population and stakeholders, as *Johan Yans* rightly points out (Chapter 4). However, the same author suggests ways to mitigate these negative perceptions;

– Second, it seems strange from an ethical point of view to expose individuals personally to a problem with the intention of raising their awareness. In the same way that we do not dump our garbage in the gardens and apartments of our fellow citizens to make them aware of sorting, it seems doubtful that we should expose them to the nuisances of mines in order to encourage them to reduce their consumption of mineral resources;

– Third, empirical statistical analyses show that countries with high domestic extraction do not consume fewer resources to support the way of life of their inhabitants – quite the contrary. The comparison of the United States and Japan presents a counter-example of lifestyle moderation through proximity to mines. If one follows the assertion described above, Japan, which has not been a major mining country since the late 19th century, should consume more material per capita than the United States, Canada or Australia, which are, on the contrary, major mining producers. However, the exact opposite is true. In 2015, Japan had a material footprint of 23 tons per capita compared to 30, 34 and 42 tons, respectively, for the three mining countries (UNEP 2016). It could be contested that these three mining countries are much less dense than Japan, which implies more consumption for the construction and maintenance of infrastructure. This is true, but does not the latter country better reflect the case of European countries? The facts are stubborn, because a study published by the PNAS (Wiedmann *et al.* 2015) shows, through a statistical analysis of 137 countries and controlling for land density, that the volume of mining per capita is positively correlated with the material footprint per capita and domestic consumption per capita. In other words, countries that extract more minerals and resources on their territory also consume more materials to sustain the lifestyles of their citizens (the study also shows this for the subset of metals);

– This naturally brings us to the fourth point. We must stop perceiving the energy transition as a supply problem that can only be solved by greater use of renewable energies (even if we do not disqualify the latter). With the exception of specific and local issues (such as chlorofluorocarbons for the hole in the ozone layer), supply-side policies alone have never succeeded in solving our global environmental problems (Dinda 2004), most of the time substituting one problem for another. It is also necessary to look at demand to cut off the pressure transfers downstream. Also, in the case of the upcoming energy transition, we must allow ourselves energy efficiency solutions and, above all, achieve greater sobriety.

One can also question the role of substitution as a natural market response to the tensions that affect mineral resource markets. Could it change the mineral resources landscape in the decades to come in a sustainable way? Intra-material substitution has always existed and will probably continue to be a predominant response to local or specific problems. Permanent magnets using rare earth metals have been replaced by copper wound rotors in a number of electric vehicle engines, following geopolitical conflicts over these materials produced almost exclusively in China. However, as *Florian Fizaine* shows (Chapter 5), this response by substitution depends not only on the technical possibilities, of course, but also on the scale of the implementation (inter-elements, inter-components, inter-system, etc.) to which are added multiple qualitative constraints of a cultural, legal, economic, physical, etc., nature. Even if this response through substitution existed and could be mobilized at a sufficient scale, the question of the implementation of "forced" substitution raises even more legitimate doubts.

It is one thing to note that substitution takes place under the effect of "natural" forces, it is quite another to trigger this mechanism through taxes and other tools at the disposal of the States. On the one hand, the objective to be achieved is sometimes totally missed or contributes to the emergence of another problem due to poor anticipation of the behavior of agents and the complex interplay of the economy. For example, Söderholm (2011) returns to the case of Sweden, a country that has taxed the extraction of gravel produced locally and intended for export (10% price increase). Initially, the measure was aimed at protecting the landscape and maintaining the availability of clean water, for which gravel reservoirs play a major role. This very indirect measure of reduction of primary extraction via the tax is theoretically quite risky as it can fail on different pitfalls or go through other channels: a weak reaction of demand to the price of the material, an increase in imports of the material (here untaxed), an increase in recycling, a substitution toward alternative untaxed materials such as crushed rock from demolition materials. In this particular example, the feedback shows that the tax has changed the behavior of the agents by pushing them more toward the substitution of crushed rock rather than toward other forms of leverage. There has been no reduction in the use of raw materials and no real increase in recycling. Moreover, crushed rock is more energy-intensive than primary gravel extraction and the production of concrete from crushed rock requires more cement, leading Söderholm to consider the policy questionable from an environmental point of view. If the aim was to reduce extraction, the economist would have advocated using regulation as operating licenses rather than economic tools such as taxes. On the other hand, mobilizing the leverage of substitution may also face social resistance. Creating an "artificial" unavailability of a resource can thus create discontent and lead to the withdrawal of the tools at its origin under social pressure (the carbon tax comes to mind). All these reasons lead us to believe that we will not profoundly modify the mineral resource landscape of our own free will via the mechanism of substitution.

In a more macroeconomic perspective, *Thierry Lefèvre* (Chapter 6) develops in his contribution the questions related to the possibility of decoupling GDP and natural resources. This question of decoupling is complex and today mobilizes a large number of researchers, particularly within the United Nations Environment Programme's (UNEP) International Resource Panel (IRP). The question of decoupling obviously refers to the tool of material efficiency, which aims to create more with less. By increasing the material productivity of our activities, we could gain in both ways: by continuing to increase GDP, while reducing our consumption of resources and the impacts left in its wake. This postulate of dematerializing the economy is an old one, notably through the concept of ephemeralization evoked by Philippe Bihouix (2019) in his latest book. We also come across it under the terms of decoupling, delinking or via the material Kuznets curve. But here again, the practical application shows poor results. Most of the time, decoupling is well below the scale effect of population and GDP per capita growth. On this point, the researcher's contribution somewhat dashes our expectations by showing that the material footprint of most industrialized countries has grown over time.

Similarly, other studies conducted worldwide (Krausmann *et al.* 2017) also temper our expectations regardless of the raw materials studied. Thus, there seems to be no exception: economic growth always outweighs material productivity. Would it be enough to increase the speed of dematerialization in order to compensate for the increase in activity? Here again, the facts contradict this idea, particularly through the example of the increase in silicon productivity in the IT sector between 1970 and 2010, which, although without precedent (a factor of 10 million), has been associated with an increase in silicon consumption of a factor of 60 over the same period! Another study on sector productivity comes to the same conclusion (Dahmus 2014): sectors that have come closer (or have reached) absolute decoupling are not characterized by a high level of material productivity but rather by a low increase in their activity (scale effect). We should therefore once again either review our objectives or look at other leverage.

Another form of leverage is deemed as highly promising, that of recycling. *Alain Geldron*'s very comprehensive contribution (Chapter 7) on the subject of metal recycling appears enlightening from several points of view. First of all, far from the sometimes blissful optimism shown by the environmental press on urban mining and the circular economy, there is a wide gap between the discourse and the empirical facts: recycling rates are still far from circularity for base metals, and are even almost zero for minor metals. Indeed, there are several fundamental differences between the extractive metal economy and the metal recycling economy, which explain why we cannot switch from one to the other without major adjustments.

First of all, the returns to scale derived from the size of the stakeholders and the volume of deposits are quite different between the two activities, clearly contributing to the domination of the first over the second. Moreover, the share of the informal sector is still very significant in the recycling economy, whereas it remains very marginal in the extractive economy, at least when we look at the volumes supplied. Second, the qualities of the materials from primary and secondary deposits differ considerably (Fizaine 2020), again with a marked disadvantage for secondary activity (dispersed deposits, highly variable and fluctuating metal concentrations, metal complexity and diversity and coexistence with carbon chains). Finally, we find the opposition between stock management and flow management as a decisive dividing line between the old extractive economy and the new secondary economy, an opposition that is not without a reminder of the same antagonistic pattern that exists in energy production. However, it is legitimate to think that the management of a flow is more complex and less flexible than that of a stock, even more so when there is significant uncertainty about the former.

To sum up, recycling research is still marked by significant gray areas. As there is now a willingness to include recycling in a comprehensive circular economy policy, together with other tools such as the reduction of primary materials (efficiency) or their reuse, it seems that we cannot simply optimize recycling procedures independently of other circular economy measures (Berlingen 2020). In this case, there is indeed a strong chance of crowding out effects between measures. The outcry over the recycling deposit measure proposed by the Secretary of State to the Minister for the

Ecological and Inclusive Transition in France, Brune Poirson, for plastic bottles is a good illustration of this. This measure is contested by local authorities, which would then be deprived of the collection and resale of this waste, for which they have already invested significant amounts in recycling infrastructures. According to the environmental associations, this project would also hamper environment preservation because the deposit for reuse is in this situation more efficient than the deposit for recycling.

Another illustration is the reduction of the precious metal content of electronic cards, for reasons of cost and efficiency, which has considerably reduced the attractiveness of recycling these cards, and also of all the minor metals that accompany them (Cui and Roven 2011; Adie *et al.* 2016). These two examples present possible incompatibilities between circular economy measures, which require careful study of the interaction effects when several measures are launched in parallel. We must also, in each situation, favor certain forms of leverage rather than generating their use across the entire circular economy.

Finally, we can see that sobriety is still the overlooked aspect of environmental policies. Often referred to in the reports of international organizations and in the wishes of companies to make their environmental balance sheet greener, moderation is not often put into practice and rarely deployed in the field or in implementing decrees. *Philippe Bihouix*'s contribution (Chapter 8) explores this possibility through a combination of objectives such as ecodesign (recyclability and product durability), "moderate use of machinery" in his terms, the resizing of activities and work on the desirability of change (highlighting the gains associated with change). Using numerous examples, he describes what could be an alternative to the search for green growth, which is confronted with complexity and often leads to rebound effects. Indeed, as we have seen in Volume 1 of this book, efficiency (providing the same economic service with less material/energy) rarely results in a decrease in consumption because this is in any case outweighed by the increase in the volume of activity. Renunciation and moderation could thus intervene where efficiency fails, by cutting off at source the primary cause of the exponential increase in the consumption of natural resources. Nevertheless, as is often forgotten, sobriety is not business-friendly and the stagnation (or even decline) in the volume of activity does not go without posing problems in terms of budget balance, debt sustainability and the financing of pension systems, notwithstanding its effects in terms of employment and unemployment. These are the questions that arise at the opening of Volume 2 of this book, which is devoted to the issues at stake and, above all, to the leverage that can provide a response to the various challenges that must be taken up in order to achieve the sustainable growth mentioned in Volume 1.

I.3. References

Adie, G.U., Sun, L., Zeng, X., Zheng, L., Osibanjo, O., Li, J. (2017). Examining the evolution of metals utilized in printed circuit boards. *Environmental Technology*, 38(13–14), 1696–1701 [Online]. Available at: doi 10.1080/09593330.2016.1237552.

Barbier, E.B. (2011). *Scarcity and Frontiers: How Economies Have Developed through Natural Resource Exploitation*. Cambridge University Press, Cambridge.

Berlingen, F. (2020). *Recyclage, le grand enfumage : comment l'économie circulaire est devenue l'alibi du jetable*. Éditions Rue de l'échiquier, Paris.

Bihouix, P. (2019). *Le bonheur était pour demain*. Le Seuil, Paris.

Cui, J. and Roven, H.J. (2011). Electronic waste. In *Waste*, Letcher, T., Vallero, D. (eds). Academic Press, Cambridge.

Dahmus, J.B. (2014). Can efficiency improvements reduce resource consumption? *Journal of Industrial Ecology*, 18(6), 883–897.

Dinda, S. (2004). Environmental Kuznets curve hypothesis: A survey. *Ecological Economics*, 49, 431–455.

Fizaine, F. (2020). The economics of recycling rate: New insights from waste electrical and electronic equipment, *Resources Policy*, 67 [Online]. Available at: https://doi.org/10.1016/j.resourpol.2020.

Krausmann, F., Wiedenhofer, D., Lauk, C., Haas, W., Tanikawa, H., Fishman, T., Miatto, A., Schandl, H., Haberl, H. (2017). Global socioeconomic material stocks rise 23-fold over the 20th century and require half of annual resource use. *PNAS*, 114(8), 1880–1885.

Pitron, G. (2018). *La Guerre des métaux rares : la face cachée de la transition énergétique et numérique*. Les Liens qui libèrent, Paris.

Söderholm, P. (2011). Taxing virgin natural resources: Lessons from aggregates taxation in Europe. *Resources, Conservation and Recycling*, 55, 911–922.

UNEP (2016). Global material flows and resource productivity. An assessment study of the UNEP International Resource Panel. United Nations Environment Programme, Paris.

Wiedmann, T.O., Schandl, H., Lenzen, M., Moran, D., Suh, S., West, J., Kanemoto, K. (2015). The material footprint of nations. *PNAS*, 112(20), 6271–6276.

Part 1

Stakes

1

Toward a New Geopolitics of Raw Materials in the Energy Transition

Emmanuel HACHE[1,2], Gondia SOKHNA SECK[1], Charlène BARNET[1], Samuel CARCANAGUE[2] and Fernanda GUEDES[1]

[1] IFP Énergies nouvelles (IFPEN), Rueil-Malmaison, France

[2] Institut de relations internationales et stratégiques (IRIS), Paris, France

1.1. Introduction

The question of access to so-called strategic materials[1] (Weil *et al.* 2009; Helbig *et al.* 2016) has always been central to economic analysis, in particular because of the random nature of the presence or absence of resources in a given territory.

While authors of the first classics, such as Malthus (1766–1834) in his *An Essay on the Principle of Population*, published in 1798, raised the question of the relationship between demographic growth and the increase in natural

1 The definition of a strategic material can be summarized as follows: it is used in many industrial sectors; it is difficult to find an adequate substitute for one or more industrial applications in the short term; the number of industrial applications using it is large and increases over time; it is used in dispersive applications with limited recycling potential; it has a high economic value; production and reserves are geographically concentrated and its production can generate major environmental externalities.

Mineral Resources Economics 2,
coordinated by Florian FIZAINE and Xavier GALIÈGUE.

food resources, it was not until the mid-19th century that the notion of dependence on foreign territories was introduced. In his famous essay *The Coal Question*, William Stanley Jevons thus took an interest in the evaluation of the United Kingdom's coal reserves to study the country's probable dependence on imported coal and the possible geopolitical consequences of its decline. In the 20th century, questions about the exhaustion of natural resources were raised by the constitution of the Club of Rome in 1968. The Club's reflections were to break with the optimism of growth without environmental externalities and expose, for the first time, the risks of systemic collapse on the basis of long-term scenarios. Thus, the report "The Limits to Growth", published in 1972, the result of simulation work by the Massachusetts Institute of Technology (MIT), was the first to propose trajectories of resource depletion, excess absorption of pollutant emissions and system collapse.

In the field of geopolitics, raw materials have most often been analyzed as a source of international conflicts, as a factor of industrial and military power or from the point of view of scarcity (Haglund 1982). Alex and Matelly (2011) proposed a more geo-economic angle by focusing on raw materials that they consider strategic by nature, due to the functioning and structure of markets and due to the involvement of a multitude of stakeholders with different objectives. These three levels of analysis remain relevant in the context of the energy transition. Indeed, investments in renewable energies (REs) are generally associated with a double dividend, since they reduce CO_2 emissions and result in a decrease in fossil fuel imports (Criqui and Mima 2012). This partial emancipation from the economic and geopolitical stakes of traditional energy security, particularly the issues of availability and accessibility, must however be analyzed in a more global logic by taking into account the materials necessary for the construction and implementation of low-carbon technologies (solar, wind, storage, electric vehicles, etc.) in national energy systems. Indeed, the necessary volumes of investment in these technologies could lead to a marked increase in the demand for materials or metals and generate major economic and geopolitical transformations in the various raw materials markets.

However, the latest UNEP-BNEF[2] report highlighted that, between 2010 and 2019, the investments already made in renewable[3] energy reached

2 The UNEP and Bloomberg New Energy Finance, https://fs-unep-centre.org/research/report, 2019.

3 The UNEP and BNEF only consider investments in the solar, wind, biomass, geothermal and small hydropower sectors.

2,600 billion dollars. The capacity of electricity generation based on renewable energy has thus increased from 414 GW to 1,650 GW in 10 years. To limit global warming to below 2°C by 2100, these investments will have to increase significantly in the coming decades.

Due to their higher material content than traditional technologies (power plants and thermal vehicles), the new needs in low-carbon technologies could thus upset the geopolitics of raw material markets. Indeed, the latter will have to integrate an additional layer of complexity related to the new relationships generated by the demand for materials in the energy transition. The geopolitics of strategic raw materials conceals a double dimension, whether one is a producer country with the main issue being the development of its resources or a consumer country for which the question of securing supplies is central. In addition to this double prism, there is also a dimension of management of resources or deposits by states and/or companies. In this context, the mineral markets are affected by issues that generally go beyond the sole prism of the economy to include geo-economics and geopolitics.

The objective of this chapter is thus to study the new economic or geopolitical dependencies that may appear in the dynamics of the global energy transition. In the first section, we present the concept of criticality established to measure the dependence of states on strategic materials and we insist on the difficult consideration of geopolitical aspects. In the second section, we then define, on the basis of our assessments of the demand for raw materials, an indicator of geological criticality, which we discuss in the light of geo-economics and geopolitics. In the third section, we examine the public policies of states to manage this new geopolitical situation.

1.2. Measuring the criticality of raw materials and geopolitical risk

1.2.1. *Criticality, strategic materials and risks*

The economic literature on the question of criticality is important, but the field of analysis remains extremely poorly defined. Indeed, while the concept of criticality can be conceptualized in a general way as an approach based on an assessment of the risks related to the production, use or end-of-life management of a raw material (Graedel and Nuss 2014), many studies focus on different disciplinary fields (economics, environment, business strategy, lifecycle analysis, etc.), use various indicators to measure criticality (R/P ratio, presence of substitutes, risk, environmental externalities, etc.) and work over multiple time horizons (from the short to medium to very long term), which make any comparison hazardous at first glance.

Behind the term "criticality", there are many risks related to the question of raw materials in the energy transition: geological risk (lack of availability of materials due to demand pressure), geopolitical risk (concentration of production in the hands of a few producing countries), economic risk (embargoes, policies of trade restrictions, market manipulation, etc.), production risk (under-investment, joint production, etc.) and environmental or social risk (emissions of pollutants related to production, health consequences, etc.).

1.2.2. *The absence of a homogeneous theoretical framework*

The starting point of any criticality study is to know from which point of view one is trying to position oneself to determine whether or not a raw material may or may not be critical. The first consideration is geographic. The assessment of criticality depends on the scale adopted, whether it is global, regional or national. There is therefore no universality of criticality. The second consideration is that of the nature of the entity consuming the raw material. In the broadest sense, this can be the economy as a whole. In a more restricted way, it will be a question of an industry, a company or even a technology. The third consideration is temporal. Indeed, technical progress, production processes and the introduction of new products on the market generate, over time, variations in the input and output of different materials on the international and national markets.

Once the framework has been defined, the studies seek to distinguish between the risks on the supply of raw materials on the one hand, and the economic and technical importance of the latter on the other hand. A third dimension has more recently been added in criticality studies: the environmental consequences linked to the production of the raw material (Graedel *et al.* 2012). Each of these three dimensions is quantified using different indices that can be aggregated according to several methods.

1.2.3. *Criticality matrices*

Criticality analyses are historically linked to the use of risk assessment matrices, which represent the intensity of a risk in two dimensions: the probability of occurrence of the risk and its severity. The first risk assessment matrix was only established in the United States in 2008 (National Research Council 2008) based on a double criterion: the economic importance of the resource and the potential risk of supply restrictions.

This methodology has been adopted by the European Commission (EC) to assess the criticality of elements in its reports published since 2011 (European Commission 2011, 2014, 2017). A major limitation of criticality analyses is the plurality of indices and their aggregation modes used to quantify the concepts of supply risk, economic importance and, when this dimension is taken into account, environmental consequences (Helbig *et al.* 2016). Among the indicators most often used in the literature are the Herfindahl Hirschman Index (HHI) of producing firms and/or producing countries (measurement of production concentration), the World Governance Index and the Global Political Risk Index (measurement of country risk), the reserves and resources still available (measurement of geological scarcity), dependence as a by-product, projections of increased demand, recycling rates of the raw material, the degree of substitutability by other materials (evaluated qualitatively) and the price of the raw material (whose volatility is sometimes preferred as a measure of market risk). The indicators are then aggregated in most cases by using weighted averages (the values of the weights vary from one study to another), by retaining only the strongest indicator, or by multiplying the indicators. Given the sensitivity of the results to methods and data, it is difficult to reach a consensus on the supply risk associated with a raw material, on the vulnerability or on the environmental consequences of the production of the material (Graedel *et al.* 2012). In this methodological framework, the ecological stakes are represented on a third axis which completes the criticality matrix. The ecological consequences of the production of a material include, on the basis of lifecycle analysis inventories, the impacts on ecosystems and human health (Graedel *et al.* 2012, 2015) (Figure 1.1).

A last gap in the existing methodology on criticality lies in the understanding of the geopolitical dimension, understood as the study of international power relations. This is generally included in the evaluation of supply risk through various indicators: the two most common are production concentration (based on the HHI) and political stability, mostly evaluated using the World Governance Index.

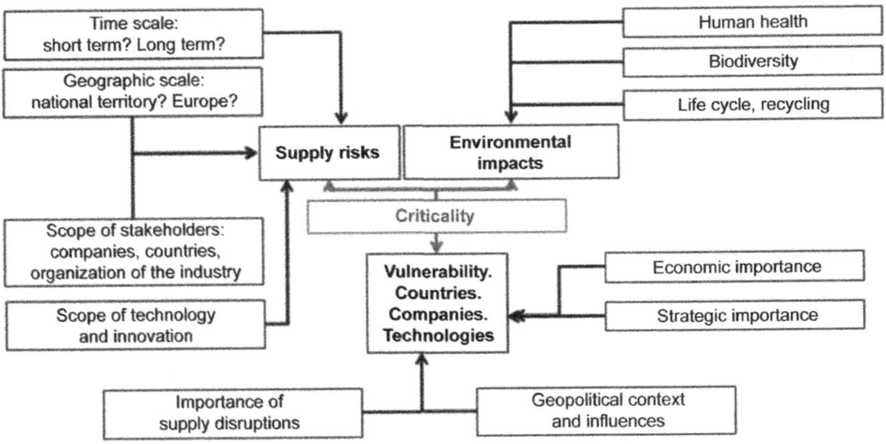

Figure 1.1. *Raw material criticality measurement factors (source: adapted from Helbig* et al. *(2016))*

Production concentration has been found to be the most commonly used indicator in criticality studies (Frenzel *et al.* 2017). The underlying geopolitical context for the assessment of production concentration is as follows: the larger the market share of an exporting country, the greater the risk that this country will be tempted to use this position to exert political or economic pressure on the importing country(ies). The geopolitics of hydrocarbons is a significant example in this regard.

However, we follow Gemechu *et al.* (2016) in their attempt to refine and quantify "geopolitical risk" by setting aside the HHI and defining the "geopolitical risk" attributed to a producer country as the sum of the risks proportionally related to its contribution to an importing country's imports (net import dependence) and its political stability. Helbig *et al.* (2016) weight this new index by including in the calculation the domestic production of the importing country, which may (or may not) compensate for a possible supply disruption. In other words, beyond market concentration, which is generic, the "geopolitical risk" is more specific since it is based on the net dependence of a client country's imports on a supplier country, an indicator that is curiously less frequently used (Frenzel *et al.* 2017). Going further, in a context of geopolitical tensions marked by "trade wars", the assessment of supply risks would be significantly enriched by taking into account the evolution of export restrictions, as suggested by Blengini *et al.* (2017). At the borderline between market and political logics, restrictions have indeed more than quadrupled between 2009 and 2014 and concern 20 of the EU's critical materials.

Metal	Year	Duration in days	Country	Effects
Cobalt	1975	–	Angola	Negligible or unknown
Cobalt	1977	60	DRC	Decline in production
Cobalt	1978	60	DRC	Suspension of production
Cobalt	1991	60	EWC	Decline in production
Titanium	1995	–	Sierra Leone	–
Copper	2004	19	DRC	–
Gold	2009	3	Argentina	Price increase
Nickel	2009	7	Papua New Guinea	Price increase, suspension of the production
Lead/zinc	2010	–	Bolivia	Development lead times
Nickel	2011	–	Philippines	Negligible or unknown
Nickel	2011	–	Indonesia	Decline in production
Nickel	2014	2	Brazil	–
Gold	2015	–	Mexico	Suspension of production
Gold	2015	30	Honduras	Suspension of production

Table 1.1. *Raw material supply disruptions due to political instability since 1967 (source: Hatayama and Tahara (2018))*

The indicator of political stability, the second most widely used indicator (Frenzel *et al.* 2017), is most often evaluated according to the World Governance Index or the Policy Perception Index (PPI). Apart from the fact that the concept of political stability is relatively vague and subject to all kinds of bias and that the relationship between good governance and political stability is not entirely linear, the causal relationship between political instability and supply risks deserves further empirical research. The case of the Democratic Republic of Congo, where the links between mineral exploitation and chronic political instability appear to be complex to say the least, illustrates this need for further investigation. Moreover, if one looks at the causes of disruption in the supply chain of 22 raw materials since 1967, only 3% are related to political instability, understood as "domestic conflict and attacks by anti-social groups" (Hatayama and Tahara 2018) (Table 1.1). The fact that political stability is a constantly used indicator in criticality studies, without being completely effective, hides in reality an analytical gap in geopolitical matters.

Indeed, a qualitative analysis of the complex and evolving nature of the relations and the balance of power between the client country (or the country in which the entity/company under consideration has its interests) and the supplier country, which is difficult to transpose into an index, is essential to assess the criticality of an ore.

The latest United States Geological Survey (USGS) methodological document on the establishment of the list of critical materials (USGS 2018) is particularly evocative from this point of view. It begins by stating:

> Most entries in the table are materials for which production concentration and net import reliance are high (typically HHI greater than 2,500 and Net Import Reliance (NIR) greater than 50 percent for either the years 2016, 2017, or both). Entries that are below the chosen threshold based on one metric or the other, but for which a case for inclusion can be made on grounds of particularly critical applications, also are included. The latter is based on the judgment of subject-matter experts of the Critical and Strategic Mineral Supply Chain (CSMSC) Subcommittee.

It is therefore understandable that the designation of final criticality will depend on a qualitative *a posteriori* analysis, based on the potentially sensitive uses of certain ores. However, the text goes on to say:

> The largest foreign suppliers of these targeted mineral commodities have been included in addition to the NIR to provide broader strategic context, which highlights that not only does the United States require foreign supplies, but that 12 out of the 26 commodities with high United States NIR are sourced primarily from China. However, high NIR should not be construed to always pose a potential supply risk. For example, three of the commodities deemed critical or near critical are primarily imported from Canada, a nation that is integrated with the United States defense industrial base.

Despite the cryptic formulation, it is understandable that criticality (or at least a form of "perceived criticality") depends largely on the identity of the supplier country, in this case China, which dominates the strategic minerals market and with which the current US presidential administration has a strained relationship. Moreover, the USGS's dependence on Canada for certain critical minerals does not explicitly appear to be a political or even economic issue. This observation leads us to question a more robust methodology in terms of criticality, which would include the political and strategic relationships between the different players in the markets studied. The interdependence relations between exporting and importing countries should also be studied, for example by developing an index of net dependence on exports for producer countries, which would weigh the index of market concentration or net dependence on imports of the client country.

Criticality is thus neither universal, timeless, nor binary (Graedel and Reck 2016). It actually varies according to the economic (commercial, technological and financial) and political (security, defense and foreign policy) interests of a state, of which it is necessarily a reflection. It is also a key to understanding the state's relations with its partners on the international scene. It thus requires the study, in parallel with the work carried out on possible geological criticality, of the different market structures and the heterogeneity of the logics of stakeholders on the different markets.

1.3. Geopolitics and geo-economics of raw materials of the energy transition

The term "geopolitics of renewable energies" is currently little used, even if some authors (Hache 2016, 2018; Scholten and Bosman 2016, 2018; Criekemans 2018) have begun to reflect on the various transformations implied by the global energy transition movement. This dynamic is not without consequences on the apprehension of the geopolitical questions posed by energy transition. In our opinion, they can be treated in two forms:

– the geological criticality of strategic materials: the multiplication of uses of critical materials generates fears of material shortages. In this context, the dynamics of consumption must therefore be considered for the year 2050;

– the fear of a cartelization of raw material markets: many markets depend on few producers or companies, which could induce the market power of certain actors on the raw material markets and thus induce *in fine* a strategic dependence on external actors.

1.3.1. *From measuring pressures on reserves to taking geopolitics into account in measuring criticality*

By reconstructing the value chains of materials essential to the energy transition – concrete/cement, cobalt, copper, lithium, nickel and rare earths – and by comparing the evolution of their respective demands up to 2050 with known resources in 2010, Hache *et al.* (2019) determined, using the TIAM-IFPEN[4] model, a criticality indicator under different scenarios (2°C and 4°C) or, in the case of concrete-cement, by post-processing the International Energy Agency (IEA) scenarios (Hache *et al.* 2020).

They then assessed the level of demand needed to meet climate objectives by 2050 and compared it to the resource assessments of the USGS to determine an indicator of criticality[5] of the different materials in the energy transition.

Metal	4 °C scenario	2 °C scenario
Cobalt	62.2%	93.6%
Copper	82.7%	96.1%
Lithium	17.1%	26%
Nickel	48.5%	56.6%
Rare earths	1.6%	3.8%

Table 1.2. *Maximum ratio of cumulative demand for materials by 2050 to identified resources 2010*

COMMENT ON TABLE 1.2. – *The figures read as follows: for cobalt, in a 4°C scenario, the cumulative global demand will represent 62.2% of the world's proven resources in 2010, compared to 93.6% in a 2°C scenario.*

Their scenarios offer differentiated results of resource pressures according to the scenarios studied. Thus, copper appears to be the most constrained metal

4 The TIAM-IFPEN model (TIMES Integrated Assessment Model), a bottom-up linear programing model, was developed at IFP Énergies nouvelles (IFPEN).

5 This concept is defined as the assessment of risks related to the production, use or end-of-life management of a raw material (Graedel and Nuss 2014).

in the energy transition dynamics, followed by cobalt. These two elements have a high criticality. By 2050, for each of these two elements, more than 90% of today's known resources would be consumed. The results for copper may seem surprising since little economic literature focuses on this subject except Vidal (2017). The sectors linked to the energy transition in the strict sense of the term (electricity grid and transport) account for a little less than a third of global copper consumption, but the other sectors taken into account in the modeling (consumer goods, industry and construction) are also expected to record significant growth rates in the coming decades.

The results observed for cobalt are also interesting, but must be put into perspective according to the type of batteries used in the transport sector. Indeed, different types of batteries have been modeled (NMC 111, NMC 622, NMC 811, LFP and NCA) and the cobalt indicator varies in the 2°C scenario from 68.5% to 93.6% depending on the penetration rate of batteries with low cobalt content. Lithium, for its part, has a low criticality since about 25% of the resources would be used at the 2050 horizon in a 2°C scenario. While the geological criticality analyses are interesting, they remain too limited because they do not allow taking into account the geo-economic or geopolitical aspects of each of these markets.

1.3.2. *Fear of cartelization or monopoly in commodity markets*

Coordinated interventions on the raw materials markets are aimed at stabilizing prices in a context of price volatility. The experience shows the difficulty of maintaining discipline within organizations over the long term. Indeed, producers have an interest in increasing their production in a context of global rationing of supply and thus positioning themselves as "free-riders". In this context, the capacity for cooperation between producers will largely depend on the internal characteristics of the market.

The creation of OPEC, in September 1960, in Baghdad by Saudi Arabia, Iran, Iraq, Kuwait and Venezuela following the rise in the demands of oil-producing countries after the World War II is the most emblematic example in the commodity markets. However, neither agricultural nor agro-industrial raw materials in the APPC (*Association des pays producteurs de café*, Association of Coffee Producing Countries), the ICCO, for cocoa, or the INRO, for rubber, nor industrial raw materials, including metals, have escaped these attempts at cartelization historically. Thus, in 1967, the CIPEC (Intergovernmental Council of Copper Exporting Countries) was created, bringing together Chile, Peru,

Zaire and Zambia, countries that accounted for 40% of world production and nearly 70% of exports. In 1975, four new countries joined CIPEC (Australia, Indonesia, Papua New Guinea and Yugoslavia); the organization disappeared in 1988, mainly due to differences between Chile and the other producer countries. In the 1970s, CIPEC was imitated by the bauxite cartel. Thus, following the creation of OPEC, the International Bauxite Association (IBA) was born. The latter was conceived progressively in the early 1960s by the government of Jamaica, then the world's second largest producer. It wished to establish between the different bauxite producers concerted actions in front of the big aluminum producers (and bauxite consumers). Jamaica imposed very high export taxes, a policy followed by other countries. However, this policy mainly benefited other IBA member countries, Australia, Guinea and Brazil, which did not pursue this policy. Indeed, these latter three countries recorded a very significant increase in their market share. Other examples such as phosphate or uranium had the same fate.

Finally, of all the cartels set up during the 19th century, after the 1930s or from the 1960s onward, only the OPEC cartel has shown, in part, its ability to control prices over short periods (Bremond *et al.* 2012). However, at the present time, two factors could revive the question of market cartelization. The first is the context of energy transition, which could gradually, with the foreseeable increase in demand, give new market power to the various raw material-producing countries or countries that contribute significantly to the raw material refining segment. This is all the more true since raw material markets are more geographically concentrated than hydrocarbon markets, both in terms of production and reserves according to the HHI [6].

The second is the return of nationalism in the raw materials sector. Indeed, since the early 2000s, with the marked rise in commodity prices and in a context of growing autocratic movements in many parts of the world, the theme of resource nationalism has made a marked return to the markets. Whether on the oil (Venezuela), gas (Ecuador) or all markets (Russia), raw materials have made a strong comeback in the logic of power of producer countries. Whether through a strategy of asserting power, or by using raw materials as a diplomatic weapon, raw materials have returned to the center of the game of international relations and the question that arises indirectly is

6 The HHI is a measure of the concentration of actors. It is calculated by summing the different market shares of each firm previously squared. HHI < 1,000 describes a competitive market, 1,000 < HHI < 1,800 describes a moderately concentrated market, and an HHI > 1,800 describes a highly concentrated market.

that of pressure on demand as a catalyst for coalition movements between countries or between companies. The examples of lithium, cobalt, copper and rare earths are interesting to analyze.

Raw materials	HHI concentration index	Main producing countries
Lithium	P: 3671	Australia (54.5%), Chile (23%), China (10%), Argentina (8%), Zimbabwe (2%), other countries (2.5%)
	R: 2963	Chile (50.5%), Australia (16%), Argentina (10%), China (6%), United States (4%), other countries (13.5%)
Cobalt	P: 5164	Congo (71.5%), Russia (4%), Australia (3.5%), Philippines (3%), Cuba (2.5%), other countries (15.5%)
	R: 2984	Congo (51%), Australia (17%), Cuba (7%), Philippines (4%), Russia (4%), other countries (17%)
Rare earth metals	P: 4317	China (63%), United States (12%), Burma (10%), Australia (10%), other countries (5%)
	R: 2163	China (37%), Brazil (18%), Vietnam (17%), Russia (10%), India (6%), other countries (12%)
Copper	P: 1118	Chile (28%), Peru (12%), China (8%), United States (6.5%), DRC (6.5%), Australia (5%), Zambia (4%), other countries (30%)
	R: 859	Chile (23%), Australia (10%), Peru (10%), Russia (7%), Mexico (6%), United States (6%) China (3%), other countries (35%)
Platinoids	P: 5412	South Africa (72%), Russia (12%), Zimbabwe (8%), Canada (4%), United States (2%), other countries (2%)
	R: 8322	South Africa (91%), Russia (6%), Zimbabwe (2%), United States (1%), other countries (< 1%)
Nickel	P: 1434	Indonesia (30%), Philippines (16%), Russia (10%), France (8%), Australia (7%), Canada (7%), China (4%), other countries (18%)
	R: 1390	Indonesia (24%), Australia (23%), Brazil (12%), Russia (8%), Cuba (7%), Philippines (6%), other countries (20%)
Oil	P: 665	United States (16%), Saudi Arabia (13%), Russia (12%), Canada (5.5%), Iran (5%), Iraq (5%), UAE (4%), other countries (39.5%)
	R: 927	Venezuela (17.5%), Saudi Arabia (17%), Canada (10%), Iran (9%), Iraq (8%), Russia (6%), Kuwait (6%), other countries (26.5%)

Table 1.3. *Geographic concentration of production (P) and reserves (R) of minerals and oil in 2019 (sources: BP statistical review (2019); USGS (2020))*

1.3.2.1. *Fear of cartelization in the lithium market?*

Historically, lithium has been used in the glass and ceramics industries, for lubricating greases and in the production of aluminum. With the development of Li-ion technologies, it has also found a strong outlet in the battery sector, for portable electronics (telephones and laptops in particular). While the use of lithium is concentrated in Asia and the United States, the mining geography is quite different. The bulk (about 90%) of raw lithium production comes from two regions: Australia and the lithium triangle in South America (Argentina, Bolivia and Chile). Despite the presence of numerous entrants in the exploration sector, the lithium market is still dominated by a small number of companies, including Chinese companies. In a context where instruments for managing price volatility are not yet available on commodity exchanges, this concentration of stakeholders leads to a high degree of uncertainty about future lithium prices. Price volatility could thus weaken new market entrants and lead to new consolidation movements (mergers and acquisitions) between players. Consolidation can also be observed with stakeholders further down the value chain. The lithium market remains a small market compared to the non-ferrous metal markets. It is characterized, in particular, by the absence of buffer stocks or regulatory stocks identified in the markets, which can lead to much more volatile price behavior and less cyclicality. The dynamics of demand can thus lead to very rapid reactions by stakeholders and high intrinsic volatility in the markets.

The historical players, American (Albemarle – formerly Rockwood – and FMC) and Chilean (SQM), extract lithium and transform it into high-purity carbonate or higher value-added lithium compounds. The two major Chinese companies, Tianqi and Ganfeng, are historically processors and have only invested in the entire value chain in recent years. In 2013, Tianqi acquired 51% of Talison's shares in Australia (the remainder is 49% owned by Albemarle). In addition to these companies, there are exploration companies, commonly referred to as juniors, whose role is to identify deposits and study the feasibility of mining them. While *the Big Three* historical players FMC, SQM and Albemarle still hold a little more than 50% of the market share today, Chinese companies were, in 2018, responsible for nearly 40% of the world's lithium supply, a large part of which is for domestic consumption.

If three groups can be distinguished, depending on whether the actors under consideration produce and/or transform lithium (producers-processors, single producers and single transformers), there is today a trend toward integration along the entire value chain and consolidation of stakeholders, in order to capture maximum value and ensure, for a given player, an increase in market share in a fast-growing market. With a limited number of players, the spot market is still underdeveloped, except in China where most transactions are carried out in the form of over-the-counter (OTC) supply contracts. However, the duration of these contracts is tending to decrease: from 1 to 3 years, the average duration has moved to 3 to 6 months, due in particular to the volatility of lithium prices. On an international scale, the lithium market is still geographically segmented between an Atlantic market (Europe and the United States being supplied mainly in South America) and a Pacific market (China having domestic production and being supplied mainly in Australia). This geographical split concerns both production and consumption and market prices.

The lithium triangle, which brings together Argentina, Bolivia and Chile, currently represents 55% of the world's resources and nearly 50% of production. Today, however, national strategies are highly differentiated. In Chile, the companies that hold the concession for the deposits are subject to exploitation quotas and limited-term leases. In Argentina, President Mauricio Macri's current economic policy in favor of foreign direct investment (FDI) is questioning the sustainability of projects in a country where foreign economic presence is a national concern that evolves with political changes. Bolivia, for its part, is a unique case in the economy of raw materials: it has the largest lithium resource in the world, located in the Uyuni Salt Flats, but does not currently produce lithium. Thus, although today there is no real rapprochement between these three great countries, national policies will be key points of understanding of the lithium market in the years to come.

In addition to the cartelization of the lithium triangle countries, there is the China problem. Indeed, Beijing imports most of the resources that it refines on its territory (including 75% spodumene from Australia and 25% brine from South America on average). As a key integrated stakeholder in the lithium sector with two major companies, Tianqi Lithium and Ganfeng Lithium Co. Beijing, have become the largest consumer of lithium and the largest producer of processed lithium. Nevertheless, China has significant resources (15% of the world's resources), even though their production cost is higher than in salt flats. We can also note the multiple Chinese acquisitions and equity investments in target markets, both in Australia and in the lithium triangle. In this context, changes in Chinese commercial policy (implementation of quotas, embargo) will have to be analyzed in light of past transformations observed in other raw material markets (rare earth metals in particular).

1.3.2.2. *A duopoly on the cobalt market?*

As an essential component of lithium-ion and nickel metal hydride (NiMH) batteries, blue metal is currently the subject of major economic stakes. World cobalt production reached 143,000 tons in 2019, of which 100,000 tons, nearly 70%, came from the Democratic Republic of Congo (DRC). In comparison, the DRC accounted for only 28% of world production in 2000 (Shedd *et al.* 2017). The next largest producers are Russia (4.3% of world production) and Australia (3.6%). Mining production is thus largely dominated by the DRC. The same type of phenomenon can be observed for refining, which is 50% controlled by China (Gulley *et al.* 2019), against 3% in 2000, the rest being mainly carried out in Finland, Belgium and Canada (Alves Dias *et al.* 2018).

Cobalt production therefore presents a supply risk with regard to three elements (Van Den Brink *et al.* 2020). First, the geographical concentration of mining and refining activities. As seen above, the DRC supplies nearly three quarters of the world's cobalt while China refines half of it. The cobalt value chain is thus split and each of the two countries has a strong influence on part of it. This concentration is also observable at the mine level: in 2016, four mining sites located in the Democratic Republic of Congo – Mutanda, Tenke Fungurume, Luiswishi and Lubumbashi Slag Hill – accounted for no less than 43% of world production (Alves Dias *et al.* 2018). The market is therefore highly concentrated at the country level for extractive activities and highly concentrated for refining, although these levels of concentration do not reach those that characterized the rare earths market in 2008 (Van Den Brink *et al.* 2020).

Second, cobalt is a by-product of copper and nickel mines. Only the Bou-Azzer mine in Morocco has as its main product the blue metal and it represents only 2% of the world supply (Alves Dias *et al.* 2018). Cobalt is therefore dependent on the price trends observed in these two associated markets. Finally, the main cobalt-producing country – the DRC – is characterized by a fragile political climate (World Bank 2020), great poverty among its population and a worrying security situation with the presence of armed groups in the east of the country (Ministère de l'Europe et des Affaires étrangères 2020). Corruption, weak governance and insufficient infrastructure[7] make the DRC an unattractive territory for businesses, as evidenced by the country's place in the "ease of doing business" ranking published by the World Bank in 2020. Out of 190 countries surveyed, the DRC ranks only 183rd in terms of ease of doing business (World Bank 2020), with New Zealand in first place and Somalia last.

It is also interesting to note that Chinese companies are very present in the DRC. Thus, in 2018, 8 of the 14 largest cobalt mines were in the hands of Chinese establishments (Farchy and Warren 2018). One of the latest examples is the acquisition by China Molybdenum of the Tenke copper and cobalt mine – the second largest in the country with 16,053 tons of cobalt produced in 2016 (Van Den Brink *et al.* 2020) – for $2.65 billion. As a result, if, for the year 2016, the share of foreign cobalt production held by Chinese entities is added to domestic production, then Chinese influence increases from 2% to 14% for cobalt extraction and from 11% to 33% for the production of cobalt intermediate products (Gulley *et al.* 2019). While Chinese dominance in the refining segment was already evident, its influence is now extending to the mining sector as well, thus moderating the weight of the DRC in the world's blue metal production.

1.3.2.3. *China and the copper market: dependence or domination?*

As the world's leading exporter, China's demand is shaping the structure and volume of international trade in raw materials, particularly in several metals of which it has become the largest consumer (iron, steel, coal, zinc, lead, tin, nickel, copper and aluminum) (Economic Commission for Latin America and the Caribbean 2018). The strategy to secure supplies implemented by China since the early 2000s has enabled it to strengthen its presence in these territories, upsetting the balance of power. The United States is therefore not happy with the growing influence that China is exerting in Africa and Latin America, particularly through the Belt and Road Initiative. The Chinese presence is perceived as a threat to democratic values and as a potential vector of anti-capitalist thinking (The White House 2017; Küblböck *et al.* 2019; Roby 2020).

Copper is no exception to the Chinese stranglehold. As the world's largest producer and consumer of electrical and electronic equipment, China

---

7 General Directorate of the Treasury, "Les défis du secteur minier en RDC", https://www.tresor economie.gouv.fr/PagesInternationales/Pages/c4657524-6a6d-4cf3-8391-9d70556e302b/files/3 fe65fde-f011-4635-9a0c-b9d0e9b41859

is at the heart of the copper value chain. However, its domestic production – 7.9% of global mining output in 2019 (USGS 2020) – is insufficient to meet its consumption and this gap is widening. Thus, the resources identified at present are not sufficient to cover the cumulative refined copper needs anticipated over the next five years (Liu *et al.* 2020). Indeed, China has many medium to small deposits, while large deposits are rarer. Added to this is the prevalence of low-grade ore deposits (Li *et al.* 2017). The identified reserves are insufficient to meet the growth in Chinese demand. In 2014, the average copper reserves were 37 years worldwide, with large disparities between countries: for example, Australian reserves are 95 years while they are only 15 years for China (Li *et al.* 2017).

To satisfy its demand, China has therefore developed its copper imports (whether in the form of ores, concentrates, blister copper, anodes or in a refined form). If the country is the leading producer of the metal in its refined form (39% of world production in 2018), it is also the main consumer of copper, and this since 2004. In 2017, China consumed 11.8 million tons of refined copper, or 50% of the total. In the same year, China's production was 8.9 million tons (ICSG 2019). The country therefore depends on imports to make up the difference: in 2017, 52% of the refined copper obtained depended on imports of copper in various forms. China's external dependence on copper ores reached 70% in 2013 (Liu *et al.* 2020). While until the late 1990s China relied mainly on domestic resources to supply its industries, it subsequently started looking for new sources of supply outside its territory (Küblböck *et al.* 2019). Securing copper supplies is therefore a fundamental issue for the emerging trade sector, whose economic model largely consists of importing raw materials from resource-rich countries, processing them and returning them as manufactured goods (Makgetla *et al.* 2019). For example, this is the form that trade relations between China and the United States have taken: the United States exports copper concentrates and scrap to China. China processes them and returns them mainly in the form of motors, generators, cables, etc. (Makgetla *et al.* 2019). Thus, one-third of Chinese copper imports are used to manufacture products for the external market (Zhang *et al.* 2017).

In order to ensure its security of supply, China has opted for increasing integration into international trade, an integration that was consecrated in 2001 by China's accession to the World Trade Organization (WTO). Initially, from 1978 to the end of the 1990s, China adopted a development strategy based on the export of low-cost products and welcomed numerous foreign investments. This model eventually ran out of steam because it did not allow China's commercial and political ambitions to be met (Küblböck *et al.* 2019). In the early 2000s, the Middle Kingdom initiated a "go out strategy" or "go global strategy", i.e. an economic and commercial development strategy encouraging Chinese companies to expand internationally. In practice, this policy is largely driven by Chinese state-owned enterprises and will result in foreign direct investment (FDI) directed mainly toward infrastructure and mineral resources in the emerging countries of Africa and Asia. These investments will help secure access to strategic raw materials for China, especially when domestic production cannot meet the demand of Chinese industries (Konijn 2014; Küblböck *et al.* 2019).

The African continent has not been the only one concerned by China's foreign policy. Between 2003 and 2013, China forged strategic partnerships with nine Latin American countries (Argentina, Brazil, Chile, Costa Rica, Ecuador, Mexico, Peru, Uruguay and Venezuela) and signed free trade agreements with three of them (Chile, Peru and Costa Rica) (Küblböck *et al.* 2019). The financing solutions proposed by the China Development Bank (CDB) and the Export and Import Bank of China (EXIM) appeared to be attractive for these countries, which were then governed by center-left governments and one of whose concerns was to break away from traditional financial institutions (Banik and Bull 2018). China has thus become the main importer of minerals and ores – iron and copper in particular – from Latin America, and $35 billion has been invested in this way by Chinese power since 2005. A total of 18 billion has been invested in copper mining projects, including 14 billion in Peru alone. Many Peruvian copper mines are now in the hands of Chinese companies: Las Bambas (bought by Minmetals in 2014, it was previously owned by Glencore), Toromocho, Galeno, Pampa del Pongo and the Rio Blanco mine (Küblböck *et al.* 2019).

China's dependence on copper imports is real. However, it has been able to diversify its sources of supply and establish deep and lasting roots in the countries of origin, whether through FDI, partnerships or free trade agreements. Today, China's strategy has borne fruit and a large part of the production apparatus of exporting countries is controlled by Chinese entities, even though the country ranks only third in terms of mining production, dominated by Chile (27.6%) and Peru (11.8%). In this way, although it is not geologically advantaged, China has been able to penetrate the copper market in terms of both mining and refined copper production.

1.3.2.4. *The issue of strategic dependence: the case of rare earth metals*

Called "vitamins of modern industry" (Balaram 2019), rare earth metals have become an essential element for many cutting-edge industries, with renewable energies (REs) and the military in the lead. Indeed, they have remarkable properties (high thermal stability, high electrical conductivity and strong magnetism) which have made it possible to achieve significant performance gains while reducing the quantity of materials consumed. They are thus incorporated in the production of solar panels and magnets for wind turbines (Judge *et al.* 2017; Dushyantha *et al.* 2020). However, while rare earth metals are in the news today, it is not so much for their qualities as for the diplomatic and commercial tensions they are used to serve. As their extraction has harmful environmental and health externalities and is economically difficult to make viable, China has taken over the global production of these elements and now holds a virtual monopoly. The crisis caused by the export restrictions implemented by the Chinese government in 2010, followed by the trade war initiated in 2018 by China and the United States, and the special attention paid to metals on this occasion have highlighted the vulnerability of developed countries that are consumers of rare earth metals. Consequently, governments – mainly the United States – have been called upon to intervene directly to support their national rare earth industry (Ahonen 2015; Eggert *et al.* 2016; Mancheri *et al.* 2019) and diversify their sources of supply.

Prior to the 1990s, China's position in the rare earths sector was still very marginal, but it rapidly became a major stakeholder in the value chain thanks to strong competitiveness enabled by cheap labor and low environmental standards (Mancheri *et al.* 2019; Seaman 2019). In 1995, China and the United States monopolized rare earth mining activities with 60.1% and 27.8% of total volumes, respectively. In 2019, global production of rare earth metals exceeded the 200,000 ton mark for the first time, with China (62% of the total), the United States (12.2%), Myanmar (10.3%) and Australia (9.9%) as the main producing countries.

Despite the increase in global production and the rise of certain countries – the United States, Australia and Myanmar – as major players in the extraction industry, China has continued to dominate the global supply chain since the mid-1990s. Indeed, if the weight of China's power in the mining sector seems to be weakening (in 2010 it accounted for almost 98% of mining production (USGS 2012)), it nevertheless retains its leadership and is also a leader in rare earth separation activities and in the production of intermediate products used in high-tech industries, a rapidly growing field (Mancheri *et al.* 2013, 2019; Mancheri 2015). However, China's rare earths sector remains mainly oriented toward low value-added activities – extraction, separation, refining and manufacture of intermediate products – and has invested less in the production of high-end, high value-added products (Mancheri 2015; Chen *et al.* 2018).

Faced with this observation, China has initiated a strategy of upgrading its industry (Dushyantha *et al.* 2020), accentuating the feeling of Chinese control over the sector (Ganguli and Cook 2018). This strategy is based, among other things, on restrictions on export volumes in order to meet its fast-growing domestic demand, on a policy of industry consolidation (Mancheri *et al.* 2019) and on dynamic scientific research. In addition to the conquest of higher value-added markets, the transformation of China's rare earth industry will also help to better control its environmental impacts. In 2010, when China introduced a series of measures restricting the supply of rare earth metals at the global level, the world became aware of the country's quasi-monopolistic position in this market and the need to diversify their sources of supply (Seaman 2019). Citing the need to limit and control the environmental impact of this industry, as well as the need to limit an over-capacity production apparatus, China introduced export quotas and increased customs tariffs. At this point, China controls more than 97% of the world's production of rare earth metals; only India and Brazil still have residual domestic production (USGS 2011). We are therefore witnessing an example of a change in trade policy directly impacting companies in consumer countries (Mancheri *et al.* 2019).

Another episode that year will be remembered: the suspension of rare earth exports to its Japanese neighbor following a diplomatic crisis related to territorial issues. These events led to the filing of a complaint with the WTO by the United States, Europe and Japan. They jointly denounced these measures which, according to them, aimed to favor Chinese companies located downstream in the value chain (World Trade Organization 2015). The intensification of the consolidation movement of the Chinese rare earth element industry was observable from 2014 onward. It mainly resulted in the consolidation of mining sector entities into six state-owned enterprises (SOEs): China Minmetals, Chinalco, Baotou Steel, Xiamen Tungsten, Ganzhou Rare Earths and Guangdong Guangsheng Rare Earths. These six SOEs have government support and favorable financing terms. Also, most of the production quotas are allocated to them (Mancheri *et al.* 2019). Similarly, the number of companies working in the field of ore separation has been reduced to 20. This restructuring of the sector reflects the Chinese government's increased control over the sector and its development. As a result, global sourcing is becoming increasingly dependent on public policies and directions taken by China (Mancheri 2015; Seaman 2019). In parallel with the reorganization movement that has begun, the transformation of the Chinese rare earth sector is perceptible in the dynamism of patent filing. Between 1950 and 2018, China filed 25,911 patents in the field of rare earth metals, compared with 9,810 for the United States, 13,920 for Japan and 7,280 for the European Union. The pace of patent filings has accelerated sharply from 2011 for China. These figures concern the entire value chain, but they do reflect the importance that China attaches to the sector and its desire to strengthen its control (Dushyantha *et al.* 2020).

The potential use of rare earth metals as a diplomatic weapon – an assumption regularly repeated in the media – must be qualified, however. In an Ifri report, John Seaman tempers the theory of a Chinese power determined to use its dominant position in the value chain to make it an instrument of first choice in the service of its diplomatic relations. For him, the actions undertaken by the Chinese government would be motivated above all by environmental considerations as well as by the desire to become a leader in cutting-edge industries requiring a certain quantity of these rare earth metals. China's position is also the subject of major evolutions: from being a supplier to the world, China has now become the main consumer of rare earth metals (Dushyantha *et al.* 2020). In 2015, China absorbed about 60% of the world's rare earth metals (Fernandez 2017). In 2018, this share rose to more than 80% (Seaman 2019). In 2020, it was expected to be 190,000 tons or 68% of the global total (Shen *et al.* 2017). Previously an exporter of unprocessed raw materials and an importer of high value-added products (Hou *et al.* 2018), China has now become the world's largest importer of rare earth metals. To accompany this economic orientation, China has developed its presence in international projects related to rare earths (Schmid 2019). In this way, China's weight as an importer and consumer of rare earth metals is expected to increase in the future, thereby changing its position and influence in the value chain (Seaman 2019).

1.4. How can we manage strategic materials supply risk?

In a globalized world, a supply disruption becomes an international phenomenon and would *de facto* require concerted actions for a multilateral management of the problem. While financial markets allow the various stakeholders to carry out hedging actions for the main non-ferrous metals, in particular on the London Metal Exchange (LME) or the Chicago Mercantile Exchange (CME) through financial contracts or options, few of these same tools are available for many strategic materials such as lithium, cobalt or rare earths. In this context, different policies can be put in place to curb supply issues.

1.4.1. *The role of public policies*

The first policy remains for states to ensure geographical and political diversification of their supplies by avoiding fragile countries and favoring stable countries. However, the diversification of supplier countries has an obvious counterpart: reduced economies of scale and higher overall import costs, due in particular to repeated and incompressible transaction costs (transport costs and contract costs). The second is the localization or relocation of activities on the national territory, which is often mentioned in the logic of securing supplies. However, while this possibility is often mentioned for many intermediate goods sectors, it is difficult to achieve in the upstream segment of the various raw materials markets because of the inexistence of material resources on national soil or because of the excessively high economic or environmental costs of production or metal refining structures. The third concerns sustainable mobility policies. These policies can in all cases reduce the pressure on the demand for materials, especially for the most critical materials (copper and cobalt). For copper in particular, the adoption of sustainable mobility policies at the international level can reduce the demand for copper in the transport sector alone by almost 20% (Seck *et al.* 2020). These policies include investment in public transport infrastructure, rationalization and optimization of transport and, more generally, discouragement of urban sprawl and changes in user behavior to promote sustainable mobility and public or non-motorized transport. Encouraging sustainable mobility policies therefore means reducing, in the medium term, imports of strategic materials.

The fourth concerns all policies relating to energy efficiency or sobriety. Public policies must put in place all the leverage to reduce pressure on resources by informing consumers about the resource impact of their purchases. More generally, energy transition policies cannot disregard a real reflection on energy sobriety, not only in its technological component of energy efficiency, but also in the collective economic and sociological dimension of slowing down demand for certain products with a high material content. The banning of disposable products, the introduction of material labeling and legislation on the crime of programed obsolescence are all essential policies to move toward a more resource-efficient world.

Finally, recycling policies are also important leverage for reducing the criticality risk for metals, especially since they can contribute to a significant environmental improvement, with a relative decrease in energy and water consumption and a reduction in greenhouse gas (GHG) emissions for metal production. For certain raw materials, Table 1.4 shows a comparison of these consumptions between a recycled raw material (waste) and a conventional extraction process (ore).

Metal	Energy (MJ/kg)		Water (m^3/ton)	
	Waste	Ore	Waste	Ore
Aluminum	10	238–925	44,181	50–600
Copper	14	31–2,040	15	40–200
Zinc	11	32–63	20	75–100
Nickel	20	130–370	20	60–320
Cobalt	20–140	140–2,100	30–100	40–2,000
Rare earths	1,000–5,000	5,500–7,200	250–1,250	1,275–1,800

Table 1.4. *Amount of water and energy used for ore extraction or reuse of waste from the mining industry (source: Sverdrup and Koca (2016))*

Recycling is thus a relevant solution in the dynamic of the energetic transition for three reasons. It provides at least a partial response to the problem of security of supply, and it helps reduce costs for industry and contributes to a relative improvement in the environmental consequences of metal production. Uncertainties about the current difficulties of the recycling industry still need to be resolved: collection costs due in particular to the geographical dispersal of the deposits, consumer behavior in waste management and the sometimes long downtime of the deposits. In addition, there is not always perfect substitutability between primary and secondary resources. This is particularly true for rare metals, which are often in very low concentrations in products.

1.4.2. *The issue of strategic stocks*

At the end of the World War I, some countries – led by the United States – became concerned about how to manage strategic resources in the new global environment. As early as 1917, the American authorities became aware of the supply difficulties and, under the aegis of the Army and Navy Munitions Board, established a list of 14 strategic materials for national defense dependent on external sources of supply and 15 critical essential materials produced on the territory in 1922. The identification of a problem relating to materials led the American authorities to decide to build up strategic stocks as early as 1939 through the Naval Appropriation Act of 1938 and the Strategic Materials Act[8] of 1939. With an initial budget of 100 million dollars, the commission identified nearly 42 strategic materials for the defense sector, particularly in view of a conflict in Europe and Asia. From then on, the United States began stockpiling materials such as rubber, manganese, tin and chromite. Between 1942 and 1946, six materials were stockpiled to meet military needs. In 1947, under the aegis of the National Security Act, the Natural Security Resources Board was created, whose objectives were to assess the needs for natural resources and to manage the acquisition and management of security stocks. In 1951, President Truman ordered the establishment of the International Materials Policy Commission, headed by William S. Paley. Its first report, entitled "Resources for Freedom" (Paley Commission 1952), noted the risks generated by external dependence on imported raw materials. The first lines are quite emblematic of the questions observed at that time (does the United States have the material means to support its civilization?) and are close to the broader problem of the lack of resources that would emerge a few years later in view of the increase in the world's population, the consequences of the massive use of pesticides on the environment, and the context of the Cold War and the partial closure of trade relations with certain countries. The confrontation with the Soviet bloc and the resulting geopolitical stakes in different parts of the world were factors that were detrimental to resource-related initiatives. The Korean War made it possible to justify all these measures since stocks of aluminum, copper and natural rubber were made available to the stakeholders during the period 1951–1956.

8 Strategic and Critical Materials Stock Piling Act June 7, 1939, ch. 190, 53 Stat. 811 (50 U.S.C. 98 *et seq.*).

In spite of numerous changes of direction after 1962, due in particular to the high cost of management[9], national stocks were built up and were used in the 1960s for cadmium (1962), lead and zinc (1964), copper (1966), quinine sulfate (1966) and nickel (1969). In 1974, during the Nixon presidency (The White House 1974), a study on the potential threat posed by the existence of foreign manipulation of the supply or price of essential non-fuel commodities fueled[10] the question of the strategic importance of materials[11]. While the reforms of 1973 and 1979 (redefinition of the concept of strategic material and transfer of competences to a new entity: the Federal Emergency Management Agency (FEMA)) should have reduced the field of the materials concerned, the election of Ronald Reagan in 1979 revived the dynamic with, as early as 1981, alerts on American vulnerability in terms of materials in its defense program. At the beginning of the 1990s, the decomposition of the USSR and the end of the Cold War led to a reconsideration of scenarios and logics linked to strategic materials. The opening up of many production zones to international trade significantly reduced the supply risk. It was not until the 2000s that questions of access to resources regained their strategic importance in the eyes of researchers and decision-makers. The open conflict on the rare earths market between China and Japan in 2011, following the Chinese territorial claims on the Senkaku Islands, confirmed the return of these issues to the heart of strategic thinking. However, the cost of managing strategic stocks remains a major barrier to this type of policy, as does the long-term identification of strategic materials that need to be stored.

1.4.3. *Foreign investment through national companies*

A relevant model from the post-World War II years, foreign direct investment (FDI) from companies in importing countries made it possible to secure supplies at lower costs. In many cases, the international relations of the time, in a post-colonial context, ensured a secure flow of raw materials to the importing countries. In many sectors, the producing countries did not have the technologies for extraction, production or marketing and the

9 President Kennedy stated that it was "staggering to observe that the storage program has accumulated $7.7 billion worth of raw materials, $3.4 billion more than the estimated wartime needs" (Snyder 1966).

10 "The President ordered a further examination of the potential threat posed by foreign manipulation of the supply or price of essential non-energy commodities" (ibid., p. 1).

11 The same reflection was carried out on the issue of energy supplies (mainly oil and petroleum products) with the creation, in 1974, of the International Energy Agency (IEA).

consumer countries imposed their own conditions, particularly in terms of tariffs. This model, which Radetzki (2008) describes as Anglo-Saxon, was weakened by the broad independence movements of the 1960s and 1970s, which in many cases led to the nationalization of production assets. However, with the wave of globalization of the 1990s, this model has regained all its logic. The example of China is particularly interesting in this context. In fact, since 2005, nearly 190 billion euros have been invested in the metals sector (Figure 1.2).

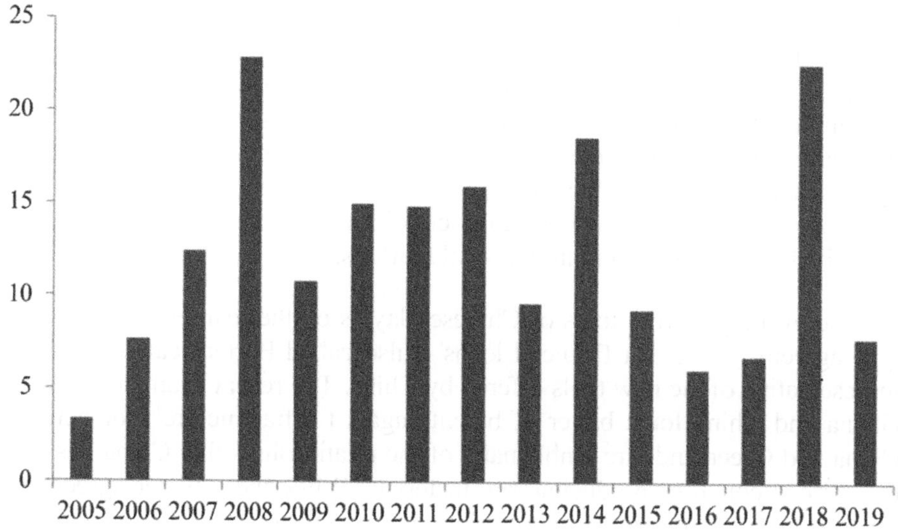

Figure 1.2. *China's foreign direct investment in the metals sector (sources: China Global Investment Tracker, The American Enterprise Institute and The Heritage Foundation)*

Another model that Radetzki (2008) describes as "*germano-allemand*" is based on long-term contracts between producer and consumer countries. These contracts, which can last nearly five years in some sectors, make it possible to secure supplies and manage price volatility. Contrary to the model based on FDI, there is no risk linked to the nationalization of assets, except to renegotiate the contract. However, this model prevented consumer countries from benefiting from low price cycles and producer countries from benefiting from high price cycles. Moreover, the financial revolution of the 1970s considerably changed the logic of long-term contracts. The 1970s thus saw two successive logics for determining raw material prices clashes. The stability relationships governed by multi-year price contracts gradually tended to disappear in favor of one year futures contracts, which were more flexible in view of market conditions and the rise of spot markets. This new situation of price volatility is prompting commodity exchanges to transform themselves to enable the various stakeholders to manage price risk.

1.4.4. *The logic of the Chinese barter*

In opposition to these approaches focused on market transactions of mineral resources, Chinese industrial actors have adopted a strategy based on what Yergin (2011) calls "constructive interdependencies". These integrate much broader cooperation than just the field of raw materials. Thus, it is not uncommon to identify the creation of multiple offers combining common political achievements, international protection agreements, bilateral trade agreements, price conditions detached from real market conditions and financing possibilities at favorable conditions in view of the real and conceivable conditions on international markets.

One of the preferred tools of Chinese players on the markets – commodity type agreements against financial loans – also called Barter deals – are fairly representative of the new tools offered by China. The recent examples between Ghana and China for a barter of bauxite against infrastructure[12], or between China and Greenland, are emblematic of the relationships that China has with the various countries producing raw materials. While these barter agreements are highly developed in the energy field, they are increasingly appearing with African or Latin American countries, rich in raw materials and lacking in infrastructure. China's interest in this type of agreement is threefold: it allows it to secure a flow of raw materials over several decades; it allows the development of its companies internationally; and Beijing obtains significant geopolitical benefits from financing infrastructures that international organizations cannot or do not want to finance. In the context of the Silk Roads undertaken by China since 2013, these agreements could see a significant increase and thus disrupt raw material markets and global geopolitics.

1.5. Conclusion: toward a new resource nationalism?

The energy transition through low-carbon technologies should lead to a marked increase in demand for certain materials (cobalt, copper, lithium,

12 https://www.alcircle.com/news/ghanas-sinohydro-barter-deal-a-model-africa-needs-to-follow-50860

nickel, rare earth elements, etc.). More than a major transformation of the geopolitics of raw materials, the dynamics observed on the markets should exacerbate the current trends. At the international level, more than 80 countries representing 25% of the world's GDP and 50% of the world's population could thus benefit from an acceleration of their production. Some emerging countries with large reserves could thus benefit from a new commodity rent, if three conditions are met. On the one hand, these countries will have to avoid the pitfall of a new Dutch disease that would curb the industrialization of other sectors of the economy. On the other hand, they will have to convert this new market power by achieving a move upmarket, which is necessary for an improvement in the quality of the national product. Finally, the ownership of a mining resource (cobalt, copper, lithium, nickel, etc.) could be a relevant asset and an interesting key to negotiation in the framework of a post-COP 21 climate agreement.

Indeed, the energy transition must not be an opportunity for these countries to allow themselves to be locked once again into a logic of dependence on raw materials. This question of dependence is today closely linked to China's place on the markets. A true "price maker" on all mineral raw materials markets, Beijing is today the main producer of strategic raw materials and has developed a refining industry such that it has more than 50% of production capacity (cobalt, copper, lithium, etc.) on its territory. It is also China that is currently deploying the most successful supply security policy, combining direct foreign investment, technological cooperation and the Barter Agreement. Like few countries before it, China has shown pragmatism in investing for nearly 30 years in the various market segments of mineral and energy raw materials. It is now present along the entire value chain for the production of mineral raw materials and at the same time is making an impressive move upmarket in low-carbon technologies (storage, electric vehicles, etc.) thanks to its Made in China 2025 plan. Since 2013, it has also been deploying the Belt and Road Initiative, a global project combining economics, diplomacy and securing supply routes.

Anxious to touch the minds and hearts of leaders, business leaders and local populations, China is thus proposing a new model combining large debt cancellations, a massive increase in aid and the construction of infrastructure, all, unlike international institutions, without imposing any particular conditions. Can this "Beijing consensus", an alternative to the "Washington consensus" of the 1980s, lead the countries producing raw materials to development (Hache 2007)?

If certain countries, Russia in particular, have been able in recent years to wave the weapon of resource nationalism, this tendency is not palpable today but could be reactivated in the years to come. Moreover, the forms of organization of the various markets (risk of cartelization, presence of oligopoly, etc.) and the lack of transparency in price formation could lead to uncertainty about the level of prices in the long term.

At the same time, the geopolitics of raw material markets could be equally disrupted in the coming years by emerging territories, notably the Arctic. For a long time sacred as a space under the seal of peace and international cooperation, the Arctic lands are today the object of renewed covetousness and growing diplomatic tensions. The melting ice has made it possible to expose the hidden, or at least inaccessible, treasures of the Arctic regions: hydrocarbon or mineral resources that have been made exploitable, areas of archeological interest, new or modified fishing resources and, finally, new maritime routes of strong economic interest (Lasserre 2021).

Finally, it is more local geopolitical elements that should increasingly interfere in this debate. The sharing of water resources is likely to be a fundamental element in the dynamics of the energy transition because production processes are water-intensive and because the main production areas for strategic materials are located in water-stressed zones.

Finally, more than a new geopolitics of raw materials, this is indeed a globalization of issues that may appear in international relations on the markets!

1.6. References

Ahonen, S., Arvanitidis, N., Auer, A., Baillet, E., Bellato, N., Binnemans, K., Blengini, G.A., Bonato, D. Brouwer, E., Brower, S. *et al.* (2015). Strengthening the European rare earths supply chain: Challenges and policy options. Technical Report, European Rare Earths Competency Network (ERECON), cea-01550114.

Alex, B. and Matelly, S. (2011). Pourquoi les matières premières sont-elles stratégiques ? *Revue internationale et stratégique*, 84, 53–60.

Alves Dias, P., Blagoeva, D., Pavel, C., Arvanitidis N. (2018). *Cobalt: Demand-Supply Balances in the Transition to Electric Mobility*. Publications Office of the European Union, EUR 29381 EN.

Balaram, V. (2019). Rare earth elements: A review of applications, occurrence, exploration, analysis, recycling, and environmental impact. *Geoscience Frontier*, 10, 1285–1303.

Banik, D. and Bull, B. (2018). Chinese engagement in Africa and Latin America: Does it matter for state capacity? *Third World Thematics: A TWQ Journal*, 3(4), 532–551.

Blengini, G.A., Nuss, P., Dewulf, J., Nita, V., Peirò, L.T., Vidal-Legaz, B., Latunussa, C., Mancini, L., Blagoeva, D., Pennington, D. *et al.* (2017). EU methodology for critical raw materials assessment: Policy needs and proposed solutions for incremental improvements. *Resources Policy*, 53, 12–19.

BP (2019). Statistical Review of World Energy [Online]. Available at: https://www.bp.com/en/global/corporate/energy-economics/statistical-review-of-world-energy.html [Accessed 26 May 2021].

Bremond, V., Hache, E., Mignon, V. (2012). Does OPEC still exist as a cartel? An empirical investigation. *Energy Economics*, 34(1), 125–131.

Chen, J., Zhu, X., Liu, G., Chen, W., Yang, D. (2018). China's rare earth dominance: The myths and the truths from an industrial ecology perspective. *Resources, Conservation & Recycling*, 132, 139–140.

Criekemans, D. (2018). Geopolitics of renewable energy game and its potential impact upon global power relations. In *The Geopolitics of Renewables*, Scholten, D. (ed.). Springer, New York.

Criqui, P. and Mima, S. (2012). European climate-energy security nexus: A model based scenario analysis. *Energy Policy*, 41, 827–842.

Dushyantha, N., Batapola, N., Ilankoon, I.M.S.K., Rohitha, S., Premasiri, R., Abeysinghe, B., Ratnayake, N., Dissanayake, K. (2020). The story of rare earth elements (REEs): Occurrences, global distribution, genesis, geology, mineralogy and global production. *Ore Geology Reviews*, 122, 103521.

Economic Commission for Latin America and the Caribbean (2018). Exploring new forms of cooperation between China and Latin America and the Caribbean [Online]. Available at: https://www.yumpu.com/en/document/read/60787635/exploring-new-forms-of-cooperation-between-china-and-latin-america-and-the-caribbean [Accessed 26 May 2021].

Eggert, R., Wadia, C., Anderson, C., Bauer, D., Fields, F., Meinert, L., Taylor, P. (2016). Rare earths: Market disruption, innovation, and global supply chains. *Annual Review of Environment and Resources*, 41, 199–222.

European Commission (2011). *Tackling the Challenges in Commodity Markets and on Raw Materials*. European Commission, Brussels.

European Commission (2014). Report on Critical Raw Materials for the EU. Report of the Ad-hoc Working Group on Defining Critical Raw Materials, European Commission, Brussels.

European Commission (2017). Report on the 2017 list of Critical Raw Materials for the EU. Report, European Commission, Brussels.

Farchy, J. and Warren, H. (2018). China has a secret weapon in the race to dominate electric cars [Online]. Available at: https://www.bloomberg.com/graphics/2018-china-cobalt/ [Accessed 26 May 2021].

Fernandez, V. (2017). Rare-earth elements market: A historical and financial perspective. *Resources Policy*, 53, 26–45.

Frenzel, M., Kullik, J., Reuter, M.A., Gutzmer, J. (2017). Raw material "criticality" – sense or nonsense? *Journal of Physics D: Applied Physics*, 50, 123002.

Ganguli, R. and Cook, D.R. (2018). Rare earths: A review of the landscape. *MRS Energy & Sustainability: A Review Journal*, 5(9).

Gemechu, E., Helbig, C., Sonnemann, G., Thorenz, A., Tuma, A. (2016). Import-based indicator for the geopolitical supply risk of raw materials in life cycle sustainability assessments: Import-based indicator for geopolitical supply risk of raw materials. *Journal of Industrial Ecology*, 20, 154–65.

Graedel, T.E. and Nuss, P. (2014). Employing considerations of criticality in product design. *The Journal of The Minerals, Metals & Materials Society (TMS)*, 66, 1–7.

Graedel, T.E. and Reck, B.K. (2016). Six years of criticality assessments: What have we learned so far? *Journal of Industrial Ecology*, 20, 692–699.

Graedel, T.E., Barr, R., Chandler, C., Chase, T., Choi, J., Christoffersen, L., Friedlander, E., Henly, C., Jun, C., Nassar, N.T. *et al.* (2012). Methodology of metal criticality determination. *Environ. Sci. Technol.*, 46, 1063–1070.

Graedel, T.E., Harper, E.M., Nassar, N.T., Nuss, P., Reck B.K. (2015). Criticality of metals and metalloids. *PNAS*, 112(14), 4257–4262.

Gulley, L., McCullough, E., Shedd, K. (2019). China's domestic and foreign influence in the global cobalt supply chain. *Resources Policy*, 62, 317–323.

Hache, E. (2007). La Chine sera-t-elle le moteur du développement de l'Afrique ? *Revue ACCOMEX*, 74, 26–33.

Hache, E. (2016). La Géopolitique des énergies renouvelables : amélioration de la sécurité énergétique et/ou nouvelles dépendances ? *Revue internationale et stratégique*, 101.

Hache, E. (2018). Do renewable energies improve energy security in the long run? *International Economics*, 156, 127–135.

Hache, E., Seck, G.S., Simoën, M., Bonnet, C., Carcanague, S. (2019). Critical raw materials and transportation sector electrification: A detailed bottom-up analysis in world transport. *Applied Energy*, 40, 6–25.

Hache, E., Seck, G.S., Simoën, M., Bonnet, C., Jabberi, A., Carcanague, S. (2020). The impact of future power generation on cement demand: An international and regional assessment based on climate scenarios. *International Economics*, 163, 114–133 [Online]. Available at: https://doi.org/10.1016/j.inteco.2020.05.002.

Haglund, D. (1982). La Nouvelle géopolitique des minéraux : une étude sur l'évolution de l'impact international des minéraux stratégiques. *Études internationales*, 13(3).

Hatayama, H. and Tahara, K. (2018). Adopting an objective approach to criticality assessment: Learning from the past. *Resources Policy*, 55, 96–102.

Helbig, C., Wietschel, L., Thorenz, A., Tuma, A. (2016). How to evaluate raw material vulnerability – An overview. *Resources Policy*, 48, 13–24.

Hou, W., Liu, H., Wang, H., Wu, F. (2018). Structure and patterns of the international rare earths trade: A complex network analysis. *Resources Policy*, 55, 133–142.

International Copper Study Group (2019). The World Copper Factbook 2019. ICSG [Online]. Available at: https://www.icsg.org/index.php/component/jdownloads/finish/170/2965 [Accessed 26 May 2021].

Judge, W.D., Xiao, Z.W., Kipouros, G.J. (2017). Application of rare earths for higher efficiencies in energy conversion. In *Rare Metal Technology 2017*, Kim, H., Alam, S., Neelameggham, N.R., Oosterhof, H., Ouchi, T., Guan, X. (eds). Springer, Berlin.

Konijn, P. (2014). Chinese Resources-For-Infrastructure (R4I) swaps: An escape from the resource curse? Document 201, South African Institute of International Affairs, African Perspectives, Global Insights, Johannesburg.

Küblböck, K., Tröster, B., Ambach, C. (2019). Going global: Chinese natural resource policies and their impacts on Latin America. Briefing document 24, Austrian Foundation for Development Research (ÖFSE), Vienna.

Lasserre, F. (2021). L'essor des activités économiques en Arctique : impact des changements climatiques et de la mondialisation. *Belgeo* [Online]. Available at: http://journals.openedition.org/belgeo/44181 [Accessed 26 May 2021].

Li, L., Pan, D., Li, B., Wu, Y., Wang, H., Gu, Y., Zuo, T. (2017). Patterns and challenges in the copper industry in China. *Resources, Conservation & Recycling*, 127, 1–7.

Liu, S., Zhang, Y., Su, Z., Lu, M., Gu F., Liu, J., Jiang, T. (2020). Recycling the domestic copper scrap to address the China's copper sustainability. *Journal of Materials Research and Technology*, 9(3), 2846–2855.

Makgetla, N., Levin, S., Mtanga, S. (2019). Moving up the copper value chain in Southern Africa [Online]. Available at: https://www.wider.unu.edu/sites/default/files/Publications/Working-paper/PDF/wp-2019-52.pdf [Accessed on 26 May 2021].

Mancheri, N.A. (2015). World trade in rare earths, Chinese export restrictions, and implications. *Resources Policy*, 46, 262–271.

Mancheri, N.A., Sundaresan, L., Chandrashekar, S. (2013). Dominating the world: China and the rare earth industry. Technical Report, International Strategy and Security Studies Programme (ISSSP), National Institute of Advanced Studies.

Mancheri, N.A., Sprecher, B., Bailey, G., Ge, G., Tukker, A. (2019). Effect of Chinese policies on rare earth supply chain resilience. *Resources, Conservation & Recycling*, 142, 101–112.

Ministère de l'Europe et des Affaires étrangères (2020). Présentation de la République démocratique du Congo [Online]. Available at: https://www.diplomatie.gouv.fr/fr/dossiers-pays/republique-democratique-du-congo/presentation-de-la-republique-democratique-du-congo/ [Accessed 15 June 2020].

National Research Council (2008). Minerals, critical materials and the U.S. Economy [Online]. Available at: https://www.nap.edu/resource/12034/critical_minerals_final.pdf [Accessed 26 May 2021].

Paley Commission (1952). Resources for freedom [Online]. Available at: https://babel.hathitrust.org/cgi/pt?id=mdp.39015028172412;view=1up;seq=9 [Accessed on 26 May 2021].

Radetzki, M. (2008). *A Handbook of Primary Commodities in the Global Economy*. Cambridge University Press, Cambridge.

Roby, V. (2020). China's growing influence in Latin America. *American Journal of Economics and Sociology*, 79(1), 233–244.

Schmid, M. (2019). Mitigating supply risks through involvement in rare earth projects: Japan's strategies and what the US can learn. *Resources Policy*, 63, 101457.

Scholten, D. (2018). *The Geopolitics of Renewables*. Springer, New York.

Scholten, D. and Bosman, R. (2016). The geopolitics of renewables; exploring the political implications of renewable energy systems. *Technological Forecasting and Social Change*, 103(C), 273–283.

Seaman, J. (2019). Rare earths and China: A review of changing criticality in the new economy [Online]. Available at: https://www.ifri.org/en/publications/notes-de-lifri/rare-earths-and-china-review-changing-criticality-new-economy [Accessed 26 May 2021].

Seck, G.S., Hache, E., Bonnet, C., Simoën, M., Carcanague, S. (2020). Copper at the crossroads: Assessment of the interactions between low-carbon energy transition and supply limitations. *Resources, Conservation & Recycling*, 163, 105072.

Shedd, K., McCullough, E., Bleiwas, D. (2017). Global trends affecting the supply security of cobalt. *Mining Engineering*, 69(12), 37–42.

Shen, L., Wu, N., Zhong, S., Li, G. (2017). Overview on China's rare earth industry restructuring and regulation reforms. *Journal of Resources and Ecology*, 8(3), 213–222.

Sverdrup, H. and Koca, D. (2016). A short description of the WORLD 6.0 model and an outline of elements of the standard parameterization. SIMRESS.

The White House (1974). National Security Study Memorandum 197: Critical imported commodities [Online]. Available at: https://fas.org/irp/offdocs/nssm-nixon/nssm_197.pdf [Accessed 26 May 2021].

The White House (2017). National Security Strategy of the United States of America. The White House, Washington.

USGS (2011). Mineral Commodity Summaries. United States Geological Survey, Reston.

USGS (2012). Mineral Commodity Summaries. United States Geological Survey, Reston.

USGS (2018). Draft critical mineral list – Summary of methodology and background information. United States Geological Survey, Reston [Online]. Available at: https://pubs.usgs.gov/of/2018/1021/ofr20181021.pdf [Accessed 26 May 2021].

USGS (2020). Mineral Commodity Summaries. United States Geological Survey, Reston [Online]. Available at: https://pubs.usgs.gov/periodicals/mcs2020/mcs2020.pdf [Accessed 26 May 2021].

Van Den Brink, S., Kleijn, R., Sprecher, B., Tukker, A. (2020). Identifying supply risks by mapping the cobalt supply chain. *Resources, Conservation & Recycling*, 156, 104743.

Vidal, O. (2017). *Mineral Resources and Energy: Future Stakes in Energy Transition*. ISTE Press, London, and Elsevier, Oxford.

Weil, M., Ziemann, S., Schebek, L. (2009). How to assess the availability of resources for new technologies? Case study: Lithium a strategic metal for emerging technologies, *Rev. Met.*, 106(12), 554–558.

World Bank (2020). Doing business 2020: Comparing business regulation in 190 economies [Online]. Available at: http://documents1.worldbank.org/curated/en/688761571934946384/pdf/Doing-Business-2020-Comparing-Business-Regulation-in-190-Economies.pdf [Accessed 26 May 2021].

World Trade Organization (2015). DS431: China – Measures related to the exportation of rare earths, tungsten and molybdenum [Online]. Available at: https://www.wto.org/english/tratop_e/dispu_e/431_432_433r_e.pdf [Accessed 26 May 2021].

Yergin, D. (2011). *The Quest: Energy, Security, and the Remaking of the Modern World*. Penguin Press, London.

Zhang, L., Chen, T., Yang, J., Cai, Z., Sheng, H., Yuan, Z., Wu, H. (2017). Characterizing copper flows in international trade of China, 1975–2015. *Science of the Total Environment*, 601–602, 1238–1246.

2

Legal Issues Regarding the Sustainable Management of Territorial and Extraterritorial Mineral Resources

Stephanie REICHE-DE VIGAN
Paris Sciences et Lettres University, France

Mineral resources are a reality of nature as well as a legal subject matter that require specific regulations to control their exploitation. The legal terminology used to refer to territorial mineral resources, i.e. mineral resources pertaining to State territory, varies from one legal system to another, and even if the concepts are identical, the associated legal regimes regarding the exploration and the exploitation of mineral resources are usually not. The notion of "mineral resources" is thus understood in this chapter in a generic sense, and refers to all non-living natural non-renewable resources, including fossil fuels and metallic and non-metallic minerals, making it possible to encompass all mineral substances regardless of their legal qualification in the legal systems considered[1].

1 While some countries have chosen for the exploitation of their territorial mineral resources to develop regulations that apply to hydrocarbons (*oil and gas law*) distinct from regulations that apply to minerals (*mining law*), this is not the case for other legal systems that apply different legal qualifications, according to the interest of the government in the mineral resource at stake,

Mineral Resources Economics 2,
coordinated by Florian FIZAINE and Xavier GALIÈGUE.

If this multiplicity of legal qualifications is far from bringing transparency and a clear understanding of the rules applicable to the exploitation of mineral resources, the fact remains that one finds great similarities among the different mineral legislations of the world when adopting a historical and comparative approach. Indeed, mineral resources law is, intrinsically, a law of exploitation of mineral resources and the question of ownership plays a key role in this. The very foundation of this branch of law is based on deciding and knowing who has the right to exploit and dispose of mineral resources for the greatest benefit of society. So far, this benefit has been found in the exploitation of mineral resources.

The same applies to extraterritorial mineral resources, that is, mineral resources that are located beyond national territories, whether they are mineral resources from the subsoil of the seabed beyond the territorial sea or mineral resources in the polar zones. As technological progress has made it possible to venture further and further away from the coastline and deeper and deeper down below sea level, states' pretensions on extraterritorial spaces and companies' claims on mineral resources therein have become more insistent. Attempts to take ownership of extraterritorial mineral resources have resulted in a national appropriation of the mineral resources of the continental shelf for the benefit of coastal states and in the collective appropriation of the mineral resources of the International Seabed Area. This area is recognized as the "common heritage of mankind" for the purpose of an equitable sharing of benefits derived from the exploitation of mineral resources "for the benefit of mankind as a whole"[2]. Yet it is for "mankind as a whole" also that states provided for a 50-year moratorium on the exploration and exploitation of mineral resources in Antarctica because of the risks that these activities could pose to this fragile ecosystem, considering that "the protection of the Antarctic environment and dependent and associated ecosystems is in the interest of mankind as a whole"[3]. Does the benefit of mankind as a whole stand in the exploitation or the non-exploitation of territorial and extraterritorial mineral resources?

Mineral resources are somewhat different from other natural resources in that, until very recently, they were considered by the legislator exclusively

---

without consideration for their geological nature or industrial method of exploitation. France, for instance, considers hydrocarbons as "mines" in a legal sense.

2 United Nations Convention on the Law of the Sea (UNCLOS) of December 10 1982, Art. 140.

3 Protocol to the Antarctic Treaty on Environmental Protection of October 4 1991, Art. 7: "Any activity relating to mineral resources, other than scientific research, shall be prohibited".

for the benefit of their exploitation – with the exception of Antarctic resources. It is necessary to envisage today, for the greatest benefit of society and in the interest of humanity as a whole, a law which no longer considers the general interest of mineral resources solely in terms of exploitation, but a law centered on the sustainable management of mineral resources. This would include – beside exploration and exploitation – the protection and conservation of resources, which would be likely to lead, for certain territorial or extraterritorial areas, to a prohibition of exploration and exploitation of mineral resources if that would adversely affect the environment or dependent and associated ecosystems and the surrounding population.

Sustainable management of mineral resources must integrate the entire lifecycle in order to involve activities that are *a priori* considered unsustainable. Although praiseworthy efforts have been made to promote greater transparency in the extractive sector through voluntary initiatives[4], current regulations do not sufficiently prevent the environmental and human adverse effects that these activities may cause, particularly in developing countries, nor do they allow for the liability of their perpetrators, whether it be the liability of the operators or the liability of the state that owns the resources for not having taken all the necessary measures to avoid them, or even that of the state of the operator's nationality for lack of control. By sustainable development we mean development that is economically efficient, socially equitable and environmentally sustainable. It must meet not only "the needs of the present without compromising the ability of future generations to meet their own needs"[5], but also the needs of local populations without compromising the ability of foreign populations to meet their own needs.

We will therefore examine the main principles of national laws (section 2.1) and of international law governing access to territorial mineral resources (section 2.2), as well as international law governing access to extraterritorial mineral resources (section 2.3), before defining a law designed to guarantee their sustainable management (section 2.4). This is the only way to ensure the future of the extractive industry, which is currently facing numerous criticisms, and also the future of humanity in a healthy environment.

4 For instance, the Extractive Industries Transparency Initiative (EITI).

5 World Commission on Environment and Development, *Our Common Future*, Oxford University Press, October 1987 (also known as the Brundtland Report).

2.1. National law regarding territorial mineral resources: the decisive issue of ownership

In domestic law, making the most of the exploitation of mineral resources has always been one of the primary concerns of states. Legislation governing the exploitation of mineral resources (or "mineral legislation") has become more complex than in ancient times (Ardaillon 1897; Mateo 2001), but continues to be based on similar mechanisms, such as concessions and royalties, with the central issue being ownership of mineral resources (or "mineral ownership"). Mineral ownership is thus "the starting point for any study of mining law" (Campbell 1957) and also the limiting factor in the the state's ability to develop rules to protect the environment from degradation associated with extractive activities. Indeed, when mineral resources are not owned by the state, the state's ability to regulate is limited by the protection afforded to private property rights when mineral resources are privately owned or by the sharing of sovereign powers over mineral resources when considering federal-type systems. The Chinese, French and American examples are particularly revealing of the legal difficulties.

2.1.1. *Ownership over mineral resources at the core of mineral law*

The distinction between land ownership and mineral ownership generally reflects the choice of the legislator as to the respective rights over mineral resources: those of the owner of the surface, those of the inventor of the deposit, those of the operator wishing to exploit it and those of the state, guardian of the general interest and of public prosperity. Mineral legislations of different countries have thus been classified into three major property regimes, identified as:

– the *state system*, by which the state reserves itself the ownership of mineral resources;

– the *accession system*, in which ownership of mineral resources is assimilated to ownership of the land;

– the *res nullius system*, in which mineral resources are declared to belong to no one.

While this classification has the advantage of simplicity, it often results in confusion between the exploration and exploitation regime or the regime of extension of ownership from surface to depth. Hence, we find common errors such as the idea that the French mineral model is based on the fact that "the riches discovered in the subsoil [are] public property, as opposed to an Anglo-Saxon model based on private property" (Tuot 2013). Nothing could be further from the truth. Indeed, the solutions developed by mineral legislation are often very subtle and, in fact, none of the above-mentioned mineral property regimes seem to exist in an absolute manner. There is, however, an important separation between legal systems in which the state reserves itself the ownership of mineral resources (or of some of them in particular) and legal systems in which mineral resources are not considered public property.

2.1.1.1. *The clarity of Chinese law: state ownership over mineral resources*

Under Chinese law, mineral resources (*kuàngchǎn zīyuán* 矿产资源) are legally defined as "the natural resources formed through geologic function, which are of value for utilization, and existing under solid, liquid or gaseous states" according to Article 2 of the Rules for Implementation of the Mineral Resources Law of the People's Republic of China of March 26 1994[6]. The varieties of mineral resources and their classification are listed in the "Detailed List of Mineral Resources" annexed to this decree. A distinction is made between 1) energy minerals, 2) metallic minerals, 3) non-metallic minerals and 4) aqueous and gaseous resources. The clarity of the Chinese regime results from the fact that the classification of mineral resources is based on the nature of the substances but also, and above all, from the fact that, all these substances being reserved by the State, no distinction needs to be made according to the holder of the property rights on mineral resources.

State ownership of mineral resources is affirmed in Article 9 of China's 1982 Constitution: "Mineral resources ... and other natural resources all belong to the State, that is, to the *entire people* ...". No resource can thus be individually owned. It is reaffirmed in Article 3 of the Rules for Implementation of the Mineral Resources Law of the People's Republic of 1994:

> All mineral resources shall be owned by the state. The state ownership of the mineral resources, either near the earth's surface or underground, shall not change with the ownership of

6 Rules for Implementation of the Mineral Resources Law of the People's Republic of China promulgated by Decree No. 152 of the State Council on March 26 1994.

the land or the right to the use of the land which the mineral resources are attached to.

State property is defined by the 1986 General Principles of Civil Law as that which "belongs to the entire people" and is "sacred and inviolable, and no organization or individual is allowed to take possession of, encroach upon, privatize, conserve or destroy it". It thus seems to correspond to a constitutionally guaranteed right (Cui *et al.* 2012) for an equitable and rational use of this property and an equitable distribution of the interests it generates (Gao 2000). State ownership would thus have as its basis and limits the duty to ensure that these resources are used in the interest of Chinese society as a whole, as seems to be stated in the 1982 Constitution of the People's Republic of China, which mentions in Article 9 that "the State shall ensure the rational use of natural resources [...]. The appropriation or degradation of natural resources by any organization or individual by any means is prohibited". In practice, however, the interests of central and local governments are sometimes far removed from those of the "whole people" (Wu 2009). Nonetheless, state ownership allows the government a great deal of freedom of action if it wishes to implement sustainable mineral resource management.

2.1.1.2. *The subtlety of French law: state administered mineral resources*

Under French law, the regime for the exploration and exploitation of mineral resources is governed by the *Code minier*. In contrast, the regime of land ownership and its extension at depth is governed by the 1804 Civil Code, and the environmental protection regime by the Environmental Code, highlighting the lack of an integrated vision of mineral resource management. The current Mineral Code, resulting from Ordinance No. 2011-91 of January 20 2011[7], makes a distinction between mineral resources according to their exploration and exploitation regime. It is by determination of the law as outlined in article L. 100-2[8] – and not according to its geological nature as article L. 100-1[9] would suggest – that a mineral

7 Order no. 2011-91 of January 20 2011, codifying the legislative part of the *Code minier*.

8 *Code minier*, Art. L. 100-2: "Any mineral or fossil substance that is not qualified by Book I of this code as a mine substance is considered as a quarry substance."

9 *Code minier*, Art. L. 100-1: "The subjection of a deposit containing mineral or fossil substances either to the legal regime of mines or to that of quarries is determined solely by the nature of the substances it contains, subject to provisions to the contrary provided for in the present code."

resource is classified either in the *mining (régime des mines)* or *quarrying* (*régime des carrières*) regimes. While French operators seem to have integrated this French legal originality, it is obvious that the French legal definition of "mine" is outdated and can be a source of error as it encompasses not only metallic minerals but also fossil fuels and non-metallic minerals.

The question of ownership of mineral resources in French law seems, to some, to have been resolved a long time ago, yet there is no legal text that mentions any ownership of mineral resources. In fact, the French *Code minier* limits itself to defining the regime of exploration and exploitation of mineral resources without ever pronouncing itself on the regime of ownership (Reiche-de Vigan 2016). According to Article L. 131-1 of the current *Code minier*, "[...] mines can only be exploited under a concession or by the State".

This French peculiarity is the result of a very specific historical context, the one in which the law of April 21 1810 was elaborated after Napoleon Bonaparte came to power. The authors of the law did not want to undermine – at least formally – the absolute character of individual property right as provided for in Article 552 of the Civil Code[10] adopted in 1804, only six years earlier, and protected by Article 17 of the Declaration of the Rights of Man and of the Citizen (DRMC) of August 26 1789[11]. Indeed, the report justifying the bill states that:

> to pronounce that *mines* are public property would have meant to annul article 552, not to modify it. This has been resolved by declaring that *mines* can only be exploited by virtue of an act of concession deliberated in the Council of State; but this act will regulate the rights of the owners of the surface on the proceeds of the conceded mines. This formal recognition of the rights of the owners is an amendment that reconciles the Civil Code and the draft bill.

10 Civil Code, Art. 552: "Ownership of the land carries with it ownership of the land beneath. The owner may make underneath all constructions and excavations that he deems appropriate, and derive from these excavations all the products that they can provide, except for modifications resulting from the laws and regulations relating to mines."

11 DRMC, Art. 17: "Since the right to Property is inviolable and sacred, no one may be deprived thereof, unless public necessity, legally ascertained, obviously requires it, and just and prior indemnity has been paid".

Therefore, it has never been asserted that mineral resources belong to the owner of the land nor that they belong to the State. The solution found in the law of 1810 by which mineral resources are administered by the State had a very precise objective, which is revealed by the words of Napoleon himself: "One can say that mines are not expressly part of the public domain, but I intend that in the end it should be so" (Pelet de la Lozère 1833). The regime defined by the law of April 21 1810 continues to be the basis of the current French mineral legislation insofar as the reforms undertaken since 1810 have not brought about any substantive change to the legal framework desired by Napoleon.

2.1.1.3. *The difficulty of U.S. law: a multiplicity of owners of property rights over mineral resources*

In U.S. law, according to the U.S. Supreme Court, "the word 'mineral' is used in so many different meanings, depending on the context, that ordinary dictionary definitions shed little light on the meaning in a given case"[12]. It is an all-encompassing concept that may embrace more or fewer resources.

Ownership of mineral resources in U.S. law is based on the common law principle *cujus est solum ejus est usque ad coelum et ad inferos,* which literally means "to whom the soil belongs owns up to heaven and down to hell", allowing the rights of the owner of the land to be extended in height and depth. The owner of the land – whether public or private – is therefore presumed to be the owner of the mineral resources under his ownership. Nevertheless, many exceptions to this general rule can be found.

Indeed, the owner of the land may sell their land while retaining ownership of the mineral resources, or sell the rights to one or all mineral resources while retaining ownership of the land, or lease one or more rights to the mineral resources. This is called *severance* of the mineral estate. Furthermore, under the *rule of capture*, so-called *fugitive* mineral resources, such as oil and gas, do not belong to anyone, not even the owner of the land, until they have been effectively captured (Daintith 2010). It is therefore necessary, since the applicable rules differ, to distinguish mineral resources within the U.S. territory: those that belong to the federal state, those that belong to the states, and mineral resources that belong to private individuals or Indian tribes.

12 Northern Pacific Railway Company v. John A. Sedrbarg (1902) 47 Law Ed 575. See also Hext v. Gill (1872), L. R. 7 Ch. 699, at p. 712; Johnstone v. Crompton Co. (1899) 2 Ch. 190.

The federal mineral estate, for example, consists of mineral resources on federal lands (onshore and offshore) or those that have been severed from a portion of land that has been sold by the federal government but for which it has retained the mineral interest. The applicable rule differs depending on the mineral resource in question. A distinction is made within the federal mineral subsoil between locatable minerals[13], leasable minerals[14] and salable minerals, each of these legal categories having a particular exploration and exploitation regime[15].

This is not the case for the private mineral estate, which corresponds to mineral resources owned by private persons, insofar as there are no general rules applicable to the exploration or exploitation of these resources that would be codified in what could be called a "mineral code". Rather, the regime varies from one state to another, with rules generally drawn from common law and related to the local history of the state and its relationship with the exploitation of mineral resources.

2.1.2. *A form of mineral ownership that may limit the government's capacity to regulate the extractive sector for environmental reasons*

Depending on whether or not the mineral resources are owned or administered by the government, the latter's ability to develop environmental standards in such a way as to govern in a sustainable way the exploration and exploitation of mineral resources will be more or less limited. The American and European examples are particularly revealing of the difficulties that may be encountered, in contrast to China, where the fight against environmental pollution has become more stringent.

---

13 These are governed by the General Mining Act of 1872 and consist of the metalliferous substances defined in section 2 of this Act: gold, silver, lead, tin, copper, zinc, uranium and "any valuable mineral deposit" not excluded by other special laws.

14 These correspond, according to the Mineral Leasing Act of 1920, which regulates research and exploitation, to liquid or gaseous hydrocarbons, coal, oil shale, potassium, phosphates and sulfur. The Federal Coal Leasing Amendments Acts of 1976 amend the Mineral Leasing Act of 1920 with respect to coal exploration and mining. Geothermal resources are also subject to the *leasing system* under the terms of the Federal Coal Leasing Act of the Geothermal Steam Act of 1970.

15 Such as gravel, sand or clay, which are governed by the Material Sales Act of 1947.

2.1.2.1. *Chinese law: toward a criminalization of environmental pollution*

An important step toward sustainable resource management was taken in 2003 with the adoption by the Central Committee of the Scientific Outlook on Development[16], which aims to give priority to the people above all. In this sense, development must be global, balanced and sustainable. It mentions that land, fresh water, energy, mineral resources and the environment have become serious constraints on economic development. It calls on the Party to make resource saving a national policy, develop a circular economy and protect the environment in order to accelerate the process of transforming Chinese society into an environment-friendly and resource-friendly society. In its March 9 2018 working report, the Supreme People's Court of China stated that 1,383 cases related to environmental damage had been brought before Chinese courts by prosecutors and 252 by non-governmental organizations over the 2013–2017 period. The following year, in its March 12 2019 working report, the Supreme Court stated, in a section entitled "China Strengthens Legal Protection of the Environment and Resources", that, in 2018 alone, 42,195 people were prosecuted for environmental and resource damage, an increase of 21% over one year. The court added that environmental pollution crimes were severely punished, with the courts closing 2,204 cases in 2018, while 251,000 other environmental and resource-related cases were tried by Chinese courts[17]. The criminalization of environmental pollution has put strong legal pressure on Chinese local governments and extractive-sector companies operating in China to comply with the requirements of environmental legislations, sometimes leading to the closure of certain companies and also to the outsourcing of certain environmental impacts related to the extractive industry out of China, because of severe environmental damages such as those caused by rare earth mining (Standaert 2019).

2.1.2.2. *European law: the exclusive competence of the Member States over the conditions of exploitation of their energy resources*

For European Member States, environmental regulation is largely a matter of European law insofar as the environment is a competence shared between the Union and the Member States, according to Article 4 of the

---

16 Decision of the CCP Central Committee on Several Important Issues for the Improvement of Socialist Market Economy Institutions, adopted on October 14 2003 (Third Plenary Meeting of the 16th CCP Central Committee).

17 http://english.court.gov.cn/reports.html.

Treaty on the Functioning of the European Union (TFEU). However, each Member State retains exclusive competence over "the conditions for exploiting its energy resources, [and] its choice between different energy sources [...]" (Art. 194 TFEU) – which corresponds, in other words, to keeping exclusive power over the exploration and exploitation of mineral resources.

As a result, the European Union does not have the competence to regulate the management of mineral resources from a sustainable development perspective. Under Article 4 of the TFEU, Member States reserve total control over essential aspects of energy policy and consequently over their climate consequences, which may call into question the Union's objective of "sustainable development of Europe based on a high level of protection and improvement of the quality of the environment" (Art. 3 of the Treaty on European Union – TEU). Already in 2007, the European Council noted that "the choice made by Member States in determining their energy mix may have effects ... on the Union's ability to achieve the three objectives of the Energy Policy for Europe" (Council of the European Union 2007), which are:

– security of supply;

– the competitiveness of European economies and the availability of affordable energy;

– environmental sustainability and the fight against global warming.

Nevertheless, in the name of the principle of loyal cooperation, provided for in Article 4 of the TEU, which requires Member States "to facilitate the achievement of the Union's tasks and refrain from any measure which could jeopardise the attainment of the Union's objectives", Member States should work to ensure that their internal regimes for the exploration and exploitation of mineral resources do not jeopardize the sustainable development of Europe based on a high level of protection and improvement of the quality of the environment. While current mineral resources legal frameworks in most Member States do not necessarily point in this direction, the fact remains that, in practice, the global share of direct extraction of mineral resources has fallen significantly in Europe, while the total volume has increased (Schandl *et al.* 2016), necessarily leading to the relocation of impacts caused by the extractive sector to other regions of the world (IPBES 2019).

2.1.2.3. *U.S. law: congressional jurisdiction in environmental matters limited for private- and state-owned lands*

In the United States, it is important to distinguish between public and private lands. On federal lands, the jurisdiction of Congress under Article IV of the Constitution has been described as "without limitations"[18], allowing it to regulate them as it sees fit. Ultimately, this possibility depends more on the political orientation of the legislature and the ecological sensitivity of the judge than on matters of competence. This is not the case for state and private lands because of the Bill of Rights adopted in 1791, which imposes many restrictions on government action. On state-owned lands, the Tenth Amendment recognizes the independence and autonomy of the states in certain reserved areas and thus limits the power of the federal government to regulate them[19]. On private lands, the Fifth Amendment gives constitutional protection to private property against action by a government (federal or state) that directly interferes with or significantly disrupts the use of its property by the owner and which is the obvious result of any environmental regulations. The possibility for the U.S. federal government to develop or adhere at the international level to environmental standards designed to ensure sustainable management of mineral resources is thus largely restricted because of the division of powers as provided for in the U.S. Constitution and the constitutional protection afforded to private property (Reiche-de Vigan 2019).

These two aspects are generally at the heart of political and legal debates in the United States relating to the development of environmental standards for the extractive sector. This was the case with the rapid development of unconventional hydrocarbons in the United States and the use of hydraulic fracturing as a recovery technique. Under President Obama's administration, the Bureau of Land Management (BLM) issued a rule in 2015 to regulate hydraulic fracturing technologies on federal and Indian lands. The purpose of this final rule was to serve "as a much-needed complement to existing regulations designed to ensure the environmentally responsible development of oil and gas resources on Federal and Indian lands, which were finalized nearly thirty years ago"[20]. This rule was struck down by the Wyoming District Court in 2016. In his decision, the judge held that the BLM had no

18 United States v. San Francisco, 310 U.S. 16, 29 (1940) and Kleppe v. New Mexico, 426 U.S. 529, 542–543.

19 United States v. Darby Lumber Co., 312 U.S. 100, 124 (1941).

20 BLM, Oil and Gas; Hydraulic Fracturing on Federal and Indian Lands; Final Rule, 80 Fed. Reg. 58, March 26th 2015 – 43 CFR Part 3160.

authority to regulate hydraulic fracturing on federal lands because Congress had not delegated its powers to the BLM[21].

Upon taking office, President Trump issued Executive Order 13783, "Promoting Energy Independence and Economic Growth", on March 28 2017 with the objective of "avoiding regulatory burdens that unnecessarily encumber energy production, constrain economic growth, and prevent job creation" and consequently with the obligation for federal agencies to:

> review existing regulations that potentially burden the development or use of domestically produced energy resources and appropriately suspend, revise, or rescind those that unduly burden the development of domestic energy resources beyond the degree necessary to protect the public interest or otherwise comply with the law.

Following this executive order, the BLM developed new rules for the extractive sector. For example, on January 31 2018, the BLM issued the Instruction Memorandum entitled "Oil and Gas Concession Reform Update – Land Use Planning and Review of Concession Boundaries"[22] to facilitate oil and gas resource development procedures on federal lands. It eliminated the use of the Master Leasing Plans introduced under the Obama administration, which were designed to prevent resource use conflicts on federal lands between the exploitation of hydrocarbon resources and other uses, and to make certain areas ineligible for concession applications for reasons such as resource[23] pre-conservation. This example of the regulation of shale gas on federal lands highlights the political rather than legal issues that arise in the United States when developing environmental standards for sustainable mineral resource management.

---

21 U.S. District Court for the District of Wyoming, Case No. 2:I5-CV-043-SWS, June 21 2016.

22 BLM, Instruction memorandum, Updating Oil and Gas Leasing Reform – Land Use Planning and Lease Parcel Reviews, IM 2018-034. U.S. Department of the Interior, January 31 2018.

23 Similarly, on February 22 2018, the BLM published a proposed rule entitled "Waste Prevention, Production Subject to Royalties, and Resource Conservation; Rescission or Revision of Certain Requirements" to revise the 2016 Final Rule developed by the BLM under the Obama Administration, which was intended to reduce natural gas waste from venting, flaring and leaks during oil and gas production activities under leases on federal and onshore Indian lands. BLM, Waste Prevention, Production Subject to Royalties, and Resource Conservation; Rescission or Revision of Certain Requirements, proposed rule. U.S. Department of the Interior, February 22, 2018.

2.2. International law regarding territorial mineral resources: the central role of state sovereignty

In international law, the management of territorial mineral resources is considered to be part of the reserved domain of states, i.e. "the field of activity in which the State, not being bound by international law, enjoys a completely discretionary competence and, consequently, must not be subject to interference by other states or international organizations" (Salmon 2001). International law is therefore, in principle, not intended to interfere with the freedom of states to regulate the exploration, exploitation and use of mineral resources as they see fit. Indeed, this freedom is based on the principle of permanent sovereignty over natural resources, recognized by the International Court of Justice as "one of the essential principles of contemporary international law"[24].

Nevertheless, given the environmental damages and human rights violations caused by certain activities in the extractive sector, the freedom of each state to manage its mineral resources must be questioned. Indeed, the principle of permanent sovereignty over mineral resources is now being challenged by the claims of indigenous peoples against the infringements of the authorizations granted by resource-owning states to explore or exploit the mineral resources present on their ancestral land and territories, in the name of their cultural identity and the relationship that unites them to the Earth.

2.2.1. *The principle of permanent sovereignty over natural resources for the benefit of international trade*

The principle of permanent sovereignty over natural resources can be defined as the right to freely have one's natural resources available in order to benefit socially and economically from the exploitation of these resources. This principle, which is theoretically absolute, is legally circumscribed by the objective of promoting international trade, which calls on States possessing mineral resources to facilitate access to them for exploitation. It is also limited in practice by the protection afforded to foreign investment, which reduces the margin of maneuver of host states in their management of mineral resources, particularly for social or environmental reasons.

---

24 ICJ, Case concerning East Timor (Portugal vs. Australia), Judgment of June 30 1995, § 29.

2.2.1.1. *A theoretically absolute principle*

The principle of permanent sovereignty over natural resources was born in contemporary international law in order to allow newly independent States to regain control over their natural resources, particularly mineral resources, over which colonial powers had control. Originally, permanent sovereignty over natural resources was the claim of a group of countries – called at the time Third World – for the promotion of their economic development, but it is in fact a claim of deserved sovereign competence associated with the very quality of state, namely the free disposal of domestic natural resources.

Resolution 523 (VI) of 1952, entitled "Integrated economic development and commercial agreements", was the first to affirm that

> under-developed countries have the right to determine freely the use of their natural resources and that they must utilize such resources in order to be in a better position to further the realization of their plans of economic development in accordance with their national interests, *and to further the expansion of the world economy.*

However, it was finally on December 14 1962, a landmark date in the development of the principle, that resolution 1803 (XVII) on "Permanent sovereignty over natural resources" was adopted, ten years after the first proposals, by 87 votes to 2, those of France and South Africa, with 12 abstentions. The principle is clearly stated: "The right of peoples and nations to permanent sovereignty over their natural wealth and resources *must be exercised in the interest of their national development and of the well-being of the people of the State concerned*" (paragraph 1), and it recognizes the freedom of the resource possessor state to prohibit exploitation if it so wishes by providing that:

> exploration, development and disposition of such resources, as well as the import of the foreign capital required for these purposes, should be in conformity with the rules and conditions which the peoples and nations freely consider to be necessary or desirable with regard to the *authorization, restriction or prohibition of such activities*. (Ibid., paragraph 2)

However, the principle is subject to many conditions which, although they allowed its adoption by the originally reluctant industrialized countries, limit its scope. The most obvious limitation is found in the protection granted to foreign investments (Barde 1981), but to this is also added the condition of exercising the principle of permanent sovereignty over natural resources, which in most of the treaties referring to it is the promotion of international trade. For example, in the African Charter on Human and Peoples' Rights adopted on June 27 1981, Article 21 provides that "[...] the free disposal of natural wealth and resources shall be without prejudice to the *obligation to promote international economic cooperation* based on mutual respect, equitable exchange, and the principles of international law". The same can be said with the Energy Charter Treaty (ECT) signed on December 17 1994[25]. Article 18 of the treaty, entitled "Sovereignty over Energy Resources", states that:

> 1. The Contracting Parties recognise state sovereignty and sovereign rights over energy resources. They reaffirm that these must be exercised in accordance with and subject to the rules of international law. 2. Without affecting the objectives of *promoting access to energy resources, and exploration and development thereof on a commercial basis* [...] 4. The Contracting Parties undertake to *facilitate access to energy resources*, inter alia, by allocating in a non-discriminatory manner on the basis of published criteria authorisations, licences, concessions and contracts to prospect and explore for or to exploit or extract energy resources.

Thus, while states have the right to freely dispose of their mineral resources, this freedom is circumscribed in most international texts by the objective of promoting international economic cooperation, which involves facilitating access to these resources for foreign operators.

2.2.1.2. *A principle limited in practice by the protection granted to foreign investments*

By entering into investment treaties to attract foreign investors to exploit the natural resources on its territory, with which it then signs contracts that are often advantageous to investors, the host state consciously or unconsciously agrees to reduce its room for maneuver in its management of mineral resources, particularly for social or environmental reasons. Indeed, the investment protection granted by treaties can even go so far as to compromise the measures taken by the state or that the state wishes to take

25 With the exception of the United States, Canada, Mexico and New Zealand.

to promote or protect human rights and the environment, as long as these measures can be qualified as "indirect expropriation" (OECD 2004; Robert-Cuendet 2010). This is the qualification used by arbitration courts that settle disputes between investors and states. The term "expropriation" corresponds, in international investment law, to the right of the state to nationalize, expropriate or transfer ownership of foreign property. It is a sovereign competence of the state (Cohen-Jonathan 1977). Expropriation requires compensation, which, according to many investment treaties, must be "prompt, adequate and effective". This formula, known as the "Hull formula", was adopted by developed countries as part of customary international law (OECD 2006), while it was rejected in the 1960s by developing countries as being only "adequate"[26].

Nevertheless, today, one can find in most investment treaties that the Hull formula is often used with only minor adjustments. In practice, capital-importing states, most of which are heavily indebted, do not have the financial means to pay compensation for the investments made by mining or oil operators, which often amount to tens of millions of dollars. This necessarily limits the ability of states to adopt social and environmental measures to regulate the extractive sector to the extent that they would be required to pay compensation plus interest because of the characterization of "indiscriminate expropriation" by the arbitration court. This treaty protection for investments is supported by a body of arbitration case law that is particularly protective of investment as well as by doctrinal articles written by authors supportive of this dispute resolution mechanism in which they are generally involved as arbitrators.

2.2.2. *A principle challenged by indigenous peoples' rights to lands, territories and resources*

State sovereignty over its territory, which includes sovereignty over the natural resources on it and therefore the right to exploit them, is challenged by indigenous peoples' right over their lands, territories and resources. Indigenous peoples are groups of individuals who preceded the existence of the state and have a close and sacred relationship with the land, both filial and spiritual, which has taken time to be recognized by the international community. While indigenous peoples' right over their lands, territories and

26 Resolution 1803 (XVII) of 1962, paragraph 4; Charter of Economic Rights and Duties of States of 1974, Art. 2.2.c.

resources has been recognized at the international level by the United Nations Declaration on the Rights of Indigenous Peoples, adopted on September 13 2007[27], in practice, the special relationship that indigenous peoples have with land and natural resources is often misunderstood by non-indigenous people and is frequently ignored by domestic legislation (Stavenhagen 2002). The rush for mineral resources has led states to give authorizations to operators to venture on or near the ancestral territories of indigenous peoples, with the aim of benefiting from the economic spin-offs, no matter how small (Auty 2006; IPBES 2019)[28], of the exploitation of resources present on their national territory. This leads to the environmental deterioration of these lands and territories and, in addition to this, to the loss of culture, traditional knowledge and livelihoods of indigenous peoples who have been dispossessed or expropriated from their lands, leading in some cases to local conflicts and forced displacement, increasing marginalization and poverty, and a decline in health among these populations, which are among the most vulnerable on the planet (OAS 2015).

2.3. International law regarding extraterritorial mineral resources: exploitation "for the benefit of mankind as a whole"

Beyond national territories, areas are, in principle, removed from territorial sovereignty. However, states have multiplied attempts to appropriate the mineral resources found there, which led to the elaboration of the 1982 United Nations Convention on the Law of the Sea (UNCLOS). The UNCLOS establishes a comprehensive legal framework governing ocean space and the resources of the sea, of the seabed and of its subsoil, for the purpose of their use and the exploitation of the resources therein. According to this convention, the mineral resources of the seabed and subsoil are subject to a degraded appropriation, which ranges from the exclusive appropriation of the mineral resources of the seabed of the territorial sea by the coastal state to the joint appropriation of the mineral resources of the international seabed area by states, and includes the finalized appropriation of the mineral resources of the continental shelf over which the coastal state

27 Resolution 61/295, annex.

28 According to IPBES: "Many countries have been unable to use mineral wealth to significantly boost their economies. By promoting the risks of local net costs, the sector is also linked to social and environmental conflicts and illegal activities, with a few large multinationals controlling vast networks of exploration sites that have committed the greatest human rights violations".

exercises sovereign rights for the purpose of exploration and exploitation. In no case has this legal framework considered prohibiting, either definitively or temporarily, the exploration and exploitation of mineral resources present on the seabed due to the risks that these activities could cause to the marine environment and marine biodiversity, as has been done for the mineral resources of the Antarctic. On the contrary, the "interest of mankind" – the counterpart of the general interest in domestic law – is considered by states to be found in the exploration and exploitation of seabed mineral resources only.

2.3.1. *The legal status of the seabed and the subsoil, determined by states' interests in the exploitation of mineral resources*

Before the exploitation of mineral resources of the seabed was made technically and economically feasible and raised the question of the legal status of this extraterritorial space, the legal approach to the uses of the seabed and subsoil was on a case-by-case basis. The question of distance from the coast was not important since the sea and the seabed beyond three miles from the coast – corresponding to the territorial sea – fell under the principle of the freedom of the seas (Grotius 1609; Gidel 1981). With the movement initiated by U.S. President Truman's 1945 executive order proclaiming that the resources on the continental shelf contiguous to the United States belonged exclusively to the country (known as the Truman Proclamation), coastal states sought to extend their sovereign powers as far as possible toward the open sea, particularly with a view to exploiting the mineral resources found there. To put an end to these territorial claims, the Third United Nations Conference on the Law of the Sea led to the adoption of the UNCLOS on December 10 1982 in Montego Bay (Jamaica) after nine years of work, the most innovative aspect of which was to declare the international seabed area "common heritage of mankind" with a view that the equitable sharing among states of the benefits derived from the development of mineral resources is done "for the benefit of mankind as a whole, irrespective of the geographical location of States, whether coastal or land-locked" (Art. 140).

2.3.1.1. *The appropriation by coastal states of the mineral resources of the territorial sea, the continental shelf and the exclusive economic zone*

At the end of World War II, energy supply became an issue of national interest and the U.S. federal government sought to ensure that most of the royalties went to the federation and not to the states. This was the goal of

President Harry S. Truman's Continental Shelf Proclamation of September 28 1945, which led to the Supreme Court of the United States in the 1950s settling various cases between the federal government and the governments of the states of California[29], Texas[30] and Louisiana[31] over revenues from the exploitation of offshore mineral resources. In response to the Supreme Court's decisions, Congress passed the Submerged Lands Act[32] (SLA) on May 22 1953, after approval by President Eisenhower who had made the subject one of his campaign arguments (Barker 1958). The SLA gave the coastal states' governments, and not the federal government, ownership of natural resources – which includes mineral resources – within three nautical miles offshore, while reaffirming, in the Outer Continental Shelf Lands Act of August 7 1953, the federal state's control and jurisdiction over natural resources on the seabed and subsoil of the continental shelf beyond the three mile limit. The U.S. Supreme Court upheld the constitutional validity of the SLA in *Alabama v. Texas* (1954)[33] by dismissing the claims of the non-coastal states of Alabama and Rhode Island.

This internal U.S. affair served as a precedent in international law. Indeed, the Truman Proclamation of 1945 led some coastal states to claim in unilateral declarations increasingly extensive[34] sea areas adjacent to their coasts. Non-coastal states, as well as the great maritime powers of that time, could not be satisfied with this multiplication of unilateral declarations. Furthermore, the international community seemed unanimous that the division of the oceans along the median lines of the coastal states should be excluded, even if the same did not necessarily apply to the seabed and subsoil.

After several failures, the need to clarify the outer limit of the continental shelf as well as the legal status of the seabed and subsoil beyond national jurisdictions was raised before the United Nations. In 1967, Malta's representative to the UN, Arvid Pardo, submitted to the UN General Assembly the question entitled: "Consideration of the question of the use of the

29 United States v. California, 332 U.S. 19 (1947).

30 United States v. Texas, 339 U.S. 707 (1950).

31 United States v. Louisiana, 339 U.S. 699 (1950).

32 Submerged Lands Act of 1953 (43 U.S.C. § 1301).

33 Alabama v. Texas, 347 U.S. 272 (1954).

34 Chile, Peru and Ecuador thus proclaimed in the Santiago Declaration of August 18 1952 their sovereignty over the continental shelf and the overlying waters up to 200 nautical miles beyond their coasts, which ultimately meant extending their territorial sea to that point.

seabed and ocean floor and the subsoil thereof on the high sea for purposes other than exclusively peaceful purposes beyond the limits of existing national jurisdiction and the exploitation of their resources in the interests of mankind", with a view to, on the one hand, prohibiting the use of the seabed for military purposes, and, on the other, establishing regulations that would allow the development of the exploitation of the riches of the sea and the seabed. This was therefore the origin of the work of the Third Conference on the Law of the Sea, which led to the 1982 UNCLOS, which it would have been more accurate to call the *United Nations Convention on the Use of Marine and Maritime Space and the Exploitation of its Resources*. Indeed, in no case is it a convention relating to the oceans in general – as its name would suggest – nor to the environmental issues in particular that this fragile ecosystem faces. On the contrary, the UNCLOS enshrined new legal spaces beyond national land territories and established corresponding legal regimes. The main objective was to regulate the exploitation of resources, particularly mineral resources, present in these different spaces (territorial sea, continental shelf, exclusive economic zone and international seabed area) in order to avoid conflicts in terms of sovereignty or ownership.

With regard to the territorial sea, UNCLOS puts an end to the question of its external limit by recognizing that "the sovereignty of the coastal State extends … to an adjacent belt of sea, described as the territorial sea; … [and] extends … to its bed and subsoil" (Art. 2) and that every state "has the right to establish the breadth of its territorial sea up to a limit not exceeding 12 nautical miles …" (Art. 3).

With regard to the continental shelf, this comprises:

> the seabed and subsoil of the submarine areas that extend beyond its territorial sea throughout the natural prolongation of its land territory to the outer edge of the continental margin, or to a distance of 200 nautical miles from the baselines from which the breadth of the territorial sea is measured where the outer edge of the continental margin does not extend up to that distance. (Art. 76, § 1)

The coastal state exercises over the continental shelf *sovereign rights* "for the purpose of exploring it and exploiting its natural resources" (Art. 77, § 1) that "do not depend on occupation, effective or notional, or on any express proclamation" (Art. 77, § 3). Thus, the legal regime governing the continental shelf as provided for in the UNCLOS is an intermediate regime between the regime of full sovereignty exercised by the State over its land territory and territorial sea and the regime of full freedom exercised by all states on the high seas. On the continental shelf, the coastal state exercises what are called sovereign rights, the ultimate purpose of which is the exploration of the continental shelf and the exploitation of its natural resources.

The exclusive economic zone (EEZ) is one of the great innovations of the Montego Bay Convention. It is an area beyond and adjacent to the territorial sea that was claimed by coastal states, particularly South American and African, who wanted to regulate the exploitation of fishery resources off their coasts[35]. The EEZ is a transitional maritime space, between the regime of sovereignty of the territorial sea and the regime of freedom of the high seas, governed by a regime of rights and jurisdiction of the coastal state and the rights and freedoms of other states. According to Article 57, the EEZ may extend up to a maximum distance of 200 nautical miles (350 km) from the baselines.

According to Article 56,

> in the exclusive economic zone, the coastal State has: (a) sovereign rights for the purpose of exploring and exploiting, conserving and managing the natural resources, whether living or non-living, of the waters superjacent to the seabed and of the seabed and its subsoil, and with regard to other activities for the economic exploitation and exploration of the zone, such as the production of energy from the water, currents and winds; (b) jurisdiction […] with regard to: the establishment and use of artificial islands, installations and structures; marine scientific research [and] the protection and preservation of the marine environment.

It adds in point 3 that "The rights set out in this article with respect to the seabed and subsoil shall be exercised in accordance with Part VI [on the continental shelf]".

The creation of this new legal space has undermined the fundamental principle of the freedom of the high seas and associated freedoms (especially fishing and exploitation of resources), which until then extended beyond the

35 The EEZ recalls in some points the Santiago Declaration of August 18 1952, in which Chile, Peru and Ecuador proclaimed their sovereignty over the continental shelf and the overlying waters up to 200 nautical miles beyond their coasts.

territorial sea, by pushing the application of this principle further offshore beyond the EEZ.

However, unlike the continental shelf, which exists independently of any proclamation, the EEZ must be declared. Thus, if a state does not declare an EEZ, the regime applicable to marine space beyond the territorial sea will be that of the high seas. The resources of the EEZ are subject to the regime of sovereign rights, i.e. the sovereign competence of coastal states, finalized because it relates to exploration and exploitation, and also, and this is quite surprising, in the context of the UNCLOS to "the protection and preservation of the marine environment" which are part of a perspective of conservation and management of natural resources. Both biological and mineral resources are thus taken into account. The coastal state thus has exclusive rights and can choose whether or not to exploit these resources.

2.3.1.2. *The so-called humanistic appropriation of the mineral resources of the international seabed area for the benefit of the economic development of states*

The greatest innovation of the UNCLOS will remain that of Article 136, which states that "the [international seabed] zone and its resources are the common heritage of mankind". Under the terms of the convention, the seabed and subsoil within the limits of national jurisdiction and the resources therein are the common heritage of mankind, the exploration and exploitation of which is in the interest of all mankind, on whose behalf the International Seabed Authority acts. The innovation lies in the desire to exclude the zone, i.e. the seabed and its subsoil beyond the limits of national jurisdiction, both from the regime of freedom and from the regime of sovereignty or sovereign rights, in order to find a "third way" for the so-called "benefit of mankind as a whole" (Reiche-de Vigan 2016). As mentioned by Mohamed Bennouna (1975),

> the interest of mankind does not transcend, as has been affirmed, the interests of individual states. On the contrary, it is the necessary reconciliation of these interests that will give the notion of mankind its real content and scope.

Such is, in fact, the solution elaborated by the Montego Bay Convention, which equates the interest of mankind as a whole with the interests of states.

Indeed, the interest of mankind lies, according to the Montego Bay Convention, in the exploration and exploitation of mineral resources of the zone, which must be encouraged by the Authority in order to ensure an equitable sharing of the financial benefits derived from such exploitation:

> in such a manner as to foster healthy development of the world economy and balanced growth of international trade, and to promote international cooperation for the over-all development of all countries, especially developing States. (Art. 150)

It should be recalled that the concept of the common heritage of mankind was born in international law in the very particular context of the Cold War, marked by the demands of the newly independent Third World states for a new world economic order and for greater equity in access to economic development (Kiss 1982).

This is, moreover, the reason why the U.S. Congress voted against the United Nations Convention on the Law of the Sea in April 1982, particularly its opposition to the regime established in Part XI for the exploitation of the seabed and subsoil beyond national jurisdiction (Clark and Meese 2007)[36]. President Reagan and his predecessors were all concerned about the independence of the United States from strategic mineral resources, present in polymetallic nodules which until then had been produced mainly in developing countries. The United States was the first to develop national regulations on the exploitation of seabed mineral resources (seabed mining), particularly beyond the continental shelf with the Deep Seabed Hard Mineral Resources Act of 1980, contrary to the regime that would be developed two years later by the UNCLOS in 1982. According to that text,

> it is the legal opinion of the United States that exploration for and commercial recovery of hard mineral resources of the deep seabed *are freedoms of the high seas* subject to a duty of reasonable regard to the interests of other states in their exercise of those and other freedoms recognized by general principles of international law.

Thus, in the 1982 Montego Bay Convention and the subsequent regulations elaborated by the International Seabed Authority, the interest of

36 Even before he became President of the United States on October 10 1978, Ronald Reagan announced in a radio program called *Ocean Mining* that "no national interest of ours could justify giving the Third World sovereign control over two-thirds of the Earth's surface" and added that "it has become apparent that the underdeveloped nations that now control the [United Nations] General Assembly are seeking 'a free ride at our expense – again'".

mankind is considered to be that of equitable global economic development through the exploitation of the mineral resources of the seabed. In no case is this development ecologically sustainable, as environmental concerns are considered in an ancillary manner and relegated to the background, unlike the regime that will be developed a few years later for exploitation of the mineral resources of the Antarctic, in which environmental concerns are at the heart of the framework. In fact, only "substantial evidence [indicating] the risk of serious harm to the marine environment" prevents prospecting activities from starting in the Area – whether for polymetallic nodules[37], polymetallic sulfides[38] or cobalt-rich ferromanganese crusts[39]. "Serious harm to the marine environment" is understood in each of the rules developed by the Authority as:

> any effect from activities in the Area on the marine environment which represents *a significant adverse change* in the marine environment determined according to the rules, regulations and procedures adopted by the Authority on the basis of internationally recognized standards and practices.

Any risk of damage to the marine environment which is not "serious" and which has not been demonstrated under "substantial evidence" is therefore insufficient to lead to the prohibition or simply restriction of the exploration or exploitation of the mineral resources in question.

2.3.2. *The legal framework for the exploitation of Antarctic mineral resources, determined by ecological considerations*

In contrast to the 1982 Montego Bay Convention which was based on the desire to exploit the mineral resources of the seabed and subsoil independently of their environmental consequences, the 1988 Convention on the Regulation of Antarctic Mineral Resource Activities (Wellington Convention) was designed to prevent any damage to the fragile ecosystem of this polar region.

The risk of escalating tensions in the Antarctic due to the willingness of some states to establish submarine bases or conduct nuclear tests has led all interested states to develop a legal framework to regulate activities in this

37 ISBA/19/C/17.

38 ISBA/16/A/12 REV. 1.

39 ISBA/18/A/11.

uninhabited territory while opting for the status quo with respect to territorial claims. The Treaty of Washington of December 1959, signed by 12 original state parties and later joined by others, led to a provisional renunciation by the state parties of any territorial claim to this frozen continent (Dupuy 1958), while deliberately excluding the question of mineral resources from the talks, as this would have made it impossible to reach agreement on the question of territorial claims.

Following the first applications for exploration permits from private companies, the advisory parties to the Antarctic Treaty developed Recommendation VII-6 entitled "Study and Discussion on the Exploitation of Antarctic Mineral Resources"[40] in 1972, which is still in effect today. In it, they recognize the particular responsibility that the Antarctic Treaty places on them, particularly for the protection of the environment and the rational use of resources; they also mention the growing interest in the possibility of exploiting the mineral resources of the Antarctic made possible by technological[41] developments, while recognizing the environmental problems that this exploitation is likely to cause. With the development of the United Nations Convention on the Law of the Sea in 1982, challenges to the Antarctic Treaty system by other UN member states increased, with some states excluded from the treaty wishing to make Antarctic mineral resources the "common heritage of mankind" along with the resources of the international seabed area with a view to sharing the benefits of such exploitation.

In order to fill the legal void left by the Antarctic Treaty, the Convention on the Regulation of Antarctic Mineral Resource Activities was signed in Wellington on June 2 1988 and elaborated a legal framework that is supposed to better preserve the Antarctic environment in the event that such activities occur, since the Antarctic Treaty did not prohibit them. The regime provided for in the Wellington Convention is particularly restrictive: extractive activities are in principle prohibited and may only be undertaken in accordance with the convention (Art. 3); no mineral resource activity shall take place *until it has been established* that the activities in question would not cause significant adverse effects on the Antarctic environment and on dependent or associated ecosystems (Art. 4); a regime of strict liability of the operator provides that they are obliged to repair any damage to the environment

40 Recommendation RCTA VII-6 (Wellington 1972) – Antarctic resources: Effects of mineral exploration, effective May 29 1975.

41 Notably those carried out in the Arctic by BP Exploration (Alaska) Inc. with the Prudhoe Bay off-shore drilling operation in Alaska, which began on April 4 1970.

resulting from their act even if they have respected their contractual obligations (Art. 8) (Couratier 1988).

However, this protective framework was applied for the reason that, while it severely restricts the carrying out of exploration and exploitation activities, it nevertheless allows it. The year 1989 saw an increase in public awareness regarding Antarctica and a willingness to preserve the continent from any kind of damage by transforming it into a peaceful natural reserve. This transformation was completed with the Protocol of the Antarctic Treaty on Environmental Protection signed in Madrid on October 4 1991 (Puissochet 1991). It makes a reversal and prohibits for a period of 50 years from the entry into force of the protocol "any activity relating to mineral resources, other than scientific research" in Antarctica, which is designated as a "natural reserve, devoted to peace and science".

Thus, while the 1982 Montego Bay Convention developed a legal framework for the exploration and exploitation of deep seabed mineral resources based on the authorization in principle of extractive activities without concern for their environmental consequences, the 1988 Wellington Convention in contrast developed a framework for Antarctic mineral resources based on the prohibition in principle of such activities, which can only be authorized if it is demonstrated that the proposed operations would not cause significant adverse effects on the environment. This is the most environmentally protective legal framework for the exploration and exploitation of mineral resources ever observed, with ecological interest taking precedence over economic interest, which should inspire any legal framework for the exploration and exploitation of mineral resources, both territorial and extraterritorial, that want to commit to sustainable development.

2.4. For a sustainable management of mineral resources

In terms of mineral resources, two opposite visions of sustainable development seem to currently be at work today.

On the one hand, there is that of a sustainable development that would actually correspond to a fair economic development. This vision is based on the exploitation of mineral resources, the benefits of which must be shared equitably among states in order to promote particularly the economic development of developing states for the purpose of international solidarity.

Such is the framework elaborated by the UNCLOS of 1982, which makes the exploration and exploitation of mineral resources in the international seabed area the principle and the prohibition of exploration and exploitation the exception, as soon as proof of serious damage to the environment is provided. As for the relevance of the exploration and exploitation of mineral resources of the seabed under national jurisdiction, this is left to the free will of coastal states.

However, the absence of consideration of all the short- and medium-term environmental impacts in authorizing extractive activities, in a context of significant biodiversity loss (IPBES 2019) likely to have considerable consequences for the future of humanity, particularly in terms of climate change, does not allow such a mineral legal framework to participate in sustainable development in the sense of an economically efficient, socially equitable and ecologically sustainable development, nor to be considered to be in the interest of "mankind as a whole" when considered as the equivalent of humanity and the community of states. Moreover, it is not obvious that the financial benefits derived by developing countries have long-term consequences on their economies or that they generate an improvement in the living conditions of local populations.

On the other hand, there is sustainable development whose primary criterion is ecological sustainability. Such is the mineral legal framework developed by the 1988 Wellington Convention on the Regulation of Antarctic Mineral Resource Activities. It considers the exploration and exploitation of mineral resources from the point of view of their environmental impact, extractive activities being authorized only when the operators can prove that the planned operations will have no negative effects on the environment other than negligible ones. We believe that this regime is the most suitable for the sustainable management of mineral resources on a global scale. Nevertheless, the current Antarctic framework, which prohibits all extractive activities, cannot be part of a global sustainable development perspective as it does not take into account the need for mineral resources to meet the needs of the present.

As with forest and water resources, which have received special attention from the international community, leading to the elaboration of strategic plans for their sustainable management, it seems clear today that sustainable management of territorial and extraterritorial mineral resources cannot be achieved without consideration of this issue by the international community, in order to bring more transparency and governance to a sector that is largely opaque, both as a cause and consequence of its lack of social acceptability. Until the international community works to understand and obtain recognition of how mineral resources contribute to sustainable development and provide economic, social, environmental and cultural benefits to present and future generations in both developed and developing countries, the sustainable management of mineral resources is, in our opinion, not feasible.

The question of the contribution of mineral resources to the achievement of the Sustainable Development Goals (SDGs)[42] has not been explored. Yet it is essential, as it is the only way to ensure the social acceptability of the extractive sector. Likewise, until the international community works to understand how each mineral resource undermines sustainable development through a lifecycle assessment that integrates all environmental (biodiversity, water, soil and climate change) and human impacts from exploration to recycling of mineral resources, and chooses to exclude certain regions of the world from exploration or exploitation because of their ecological fragility, the sustainable management of mineral resources will not be possible.

Sustainable management of mineral resources also requires the implementation of the principle of due diligence (Reiche-de Vigan 2018a) in order to increase the responsibility of the extractive sector and fight against impunity in cases of damage. *Due diligence* obligations are obligations of conduct that do not imply the need to prove that material damage has been caused but only that the perpetrator has breached their obligation by failing to take all the measures that should have been taken. Due diligence may play an important role in encouraging the international responsibility of the resource possessor state with regard to authorizing extractive activities in environmentally sensitive areas or failing to put in place the necessary standards to ensure responsible operations. Due diligence is also intended to play a significant role in triggering the responsibility of the state of the nationality of the operator undertaking mineral activities abroad whose activities have resulted in environmental damage or human rights violations, for failure to control its nationals. Finally, due diligence is intended to be invoked against the operator itself in the name of its duty of vigilance.

Only then will the sustainable management of territorial and extraterritorial mineral resources based on economically efficient, socially

42 The SDGs are the 17 goals set by UN member states in Agenda 2030, adopted in September 2015. UN, "Transforming our world: The 2030 Agenda for Sustainable development", A/RES/70/1.

equitable and ecologically sustainable development that meets the needs of local populations without compromising the ability of foreign populations to meet their own, as well as meeting the needs of the present without compromising the ability of future generations to meet their own, be possible.

2.5. References

Aguillon, L. (1886). *Législation des mines française et étrangère*, volume 1. Baudry et Cie, Paris.

Ardaillon, E. (1897). *Les Mines du Laurion dans l'Antiquité*. Bibliothèque des Écoles françaises d'Athènes et de Rome, Le Bigot frères, Paris.

Auty, R. (2006). Mining enclave to economic catalyst: Large mineral projects in developing countries. *The Brown Journal of World Affairs*, 13(1), 135–145.

Barde, J. (1981). *La Notion de droits acquis en droit international public*. Publications universitaires de Paris, Paris.

Barker, J. (1958). The Offshore Oil Controversy Since 1953. *Wisconsin Law Review*, Madison, WI.

Bennouna, M. (1975). Le fond des mers : de "l'héritage commun" à la "querelle des héritiers". *Revue iranienne des relations internationales*, 5–6(1975–1976), 130.

Campbell, N. (1957). Principles of mineral ownership in the civil law and common law systems. *Tulane Law Review*, 31, 304.

Clark, W.P. and Meese, E. (2007). Reagan and the law of the sea. *Wall Street Journal*, 8 October.

Cohen-Jonathan, G. (1977). L'Arbitrage Texaco-Calasiatic contre Gouvernement Libyen ; décision au fond du 19 janvier 1977. *A.F.D.I.*, 23, 452–479.

Combacau, J. and Sur, S. (2014). *Droit international public*, 11th edition. LGDJ, Paris.

Council of the European Union (2007). Conclusion de la présidence du Conseil européen de Bruxelles des 8 et 9 mars 2007, 7224/1/07. Conseil de l'Union Européenne, Brussels.

Couratier, J. (1988). La Convention sur la réglementation des activités relatives aux ressources minérales de l'Antarctique (Wellington – 2 juin 1988). *A.F.D.I.*, 34, 772–774.

Cui, J.Y., Peng, C.X., Dai, M.Y. (2012). Study on natural resources property legal system. *Law Press China*.

Daintith, T. (2010). *Finders Keepers? How the Law of Capture Shaped the World Oil Industry*. RFF Press, Washington.

Dupuy, R.-J. (1958). Le Statut de l'Antarctique. *A.F.D.I.*, 4, 196–229.

Gao, F.P. (2000). Legal connotation of public and private ownership. *Tsinghua Law Review*, 3rd edition. Tsinghua Law School, Beijing.

Gidel, G. (1981). *Le Droit international public de la Mer – le temps de paix*, volume I. Librairie Edouard Duchemin, Paris.

Grotius, H. (1609). *La Liberté des mers – Mare Liberum*. Éditions Panthéon-Assas, Paris.

Héron de Villefosse, A.-M. (1810). *De la richesse minérale, considérations sur les mines, usines et salines des différens États, et particulièrement du Royaume de Westphalie, pris pour terme de comparaison*, volume 1. Berger-Levrault, Paris.

IPBES (2019). Global assessment on biodiversity and ecosystem services [Online]. Available at: https://ipbes.net/global-assessment.

Kiss, A. (1982). La Notion de patrimoine commun de l'humanité. *R.C.A.D.I.*, 175(II), 99 sq.

Mateo, A. (2001). Observaciones sobre el régimen jurídico de la minería en tierras públicas en época romana. University of Santiago de Compostela, Santiago de Compostela.

OAS (2015). Indigenous peoples, Afro-descendent communities, and natural resources: Human rights protection in the context of extraction, exploitation, and development activities [Online]. Available at: http://www.oas.org/en/iachr/reports/pdfs/ExtractiveIndustries2016.pdf.

OECD (2004). *L'"expropriation indirecte" et le "droit de réglementer" dans le droit international de l'investissement*. Éditions OCDE, Paris.

OECD (2006). *Droit international de l'investissement : un domaine en mouvement complément aux perspectives de l'investissement international*. Éditions OCDE, Paris.

Pelet de la Lozère, J. (1833). *Opinions de Napoléon sur divers sujets de politique et d'administration recueillies par un membre de son Conseil d'État*. Frimin Didot Frères, Paris.

Puissochet, J.-P. (1991). Le Protocole au traité sur l'Antarctique, relatif à la protection de l'environnement. *A.F.D.I.*, 37(1), 755–773.

Reiche-de Vigan, S. (2016). Le Droit et l'espace souterrain. Enjeux de propriété et de souveraineté en droit international et comparé. PhD Thesis, Jean Moulin University Lyon 3, Lyon.

Reiche-de Vigan, S. (2018a). *Due diligence* et droit international de l'environnement : le devoir de prendre soin, contrepartie du droit à un environnement sain ? *Due diligence des États, des entreprises et des individus, conférence pour les 70 ans de la Déclaration universelle des droits de l'Homme*. Court of Cassation, Paris.

Reiche-de Vigan, S. (2018b). Le droit des peuples autochtones sur leurs ressources naturelles : le principe de souveraineté permanente, seul rempart contre la désacralisation du droit de propriété ? *Actes du colloque international "La protection des minorités ethniques, culturelles et religieuses : un enjeu de paix et de sécurité internationales à l'heure de la mondialisation", Études interculturelles*, 12.

Reiche-de Vigan, S. (2019). Protection de l'environnement et changement climatique : examen comparé de l'équilibre des compétences dans l'Union européenne et aux États-Unis. *Revue de l'Union européenne*.

Robert-Cuendet, S. (2010). *Droits de l'investisseur étranger et protection de l'environnement – contribution à l'analyse de l'expropriation indirecte*. Martinus Nijhoff Publishers, Leyde.

Salmon, J. (2001). *Dictionnaire de droit international public*. Bruylant, Brussels.

Schandl, H., Fischer-Kowalski, M., West, J., Giljum, S., Dittrich, M., Eisenmenger, N., Geschke, A., Lieber, M., Wireland, H., Schaffartzik, A. *et al.* (2016). Global material flows and resource productivity. *An Assessment Study of the UNEP International Resource Panel*. United Nations Environment Programme, Paris.

Standaert, M. (2019). China wrestles with the toxic aftermath of rare earth mining [Online]. Available at: https://e360.yale.edu/features/china-wrestles-with-the-toxic-aftermath-of-rare-earth-mining.

Stavenhagen, R. (2002). Report of the Special Rapporteur on the situation of human rights and fundamental freedoms of indigenous people [Online]. Available at: https://digitallibrary.un.org/record/665137?ln=fr.

Tuot, T. (2013). Projet de réforme du Code minier, note de synthèse des propositions de Thierry Tuot [Online]. Available at: https://serdeaut.univ-paris1.fr/fileadmin/cerdeau/son/2_-_Synthèse_des_propositions_TTuot_-_Dossier_de_presse_01.pdf.

Wu, Y. (2009). Researching on the reform of Chinese mineral resources property rights institution: From the angle of central and local government decentralization. Thesis, Southwestern University of Finance and Economics, Chengdu.

3

Mining and Societies

Michel DESHAIES
LOTERR, University of Lorraine, Nancy, France

3.1. Introduction

Mineral wealth has long played a major role in the spread of settlement on the Earth's surface. From the mountains of Europe to the deserts of the American West, from the icy solitudes of Siberia to the frozen desolate areas of Canada's far north, many areas that were initially deserted (or almost deserted) only began to be more densely populated after the discovery of metal deposits, especially precious metals such as gold and silver. Very early on, the exploitation of minerals also had a significant impact on the landscapes and environment of mining regions. In the industrial age, these same metal resources, and even more so energy deposits such as coal, served as a support for industrialization, multiplying the needs for the workforce. The evolutions then contrasted strongly, depending on whether the industry developed on the deposit, as was the rule for coal, or outside the deposit, as was more often the case for metal deposits. The result today is considerable contrasts in settlement between the densely populated coalfields and the situation of many metal-bearing deposits with low population densities.

Starting in the second half of the 20th century, the increase in mid-century exploitation in the world took place mainly in new regions located in Asia, Australia, Latin America and Africa, while European deposits and those in the formerly industrialized part of North America saw their production decline, before a complete shutdown at the beginning of the 21st century in Western Europe.

Mineral Resources Economics 2,
coordinated by Florian FIZAINE and Xavier GALIÈGUE.

These new mining regions are characterized by the gigantic forms of exploitation where great discoveries dominate, causing in a few years a disruption of the landscape, while strong mechanization limits the population. Developing in semi-desert or agricultural regions, populated by populations with no mining tradition, these new exploitations often arouse resolute opposition, supported by NGOs that contest the environmental risks of mining. At a time when the need for coal and minerals of all kinds has never been greater, mining, which is reaching dimensions unprecedented in the past, is increasingly under the spotlight for its environmental impact.

3.2. Mines as a factor of settlement and landscape transformation

3.2.1. *Mining and the population*

The role of mining in the spread of settlement has varied greatly depending on geological and geographic conditions. A distinction must be made between regions that were already densely populated at the time the deposits were mined and those that were only populated following the discovery of mineral wealth. The first case is generally illustrated by European coal deposits, located in sedimentary basins where the quality of the soil had allowed early agricultural exploitation, sometimes as early as the Neolithic period, or at the latest in the Middle Ages. The long string of coal or lignite basins that runs from northern France to Silesia, via the Belgian Borinage, the Dutch Limburg and the Ruhr in Germany, provides a demonstrative example, since these are also fertile agricultural regions, thanks to the loess[1] that forms the substrate of their soil.

The second case concerns a large part of the metal deposits. Indeed, the processes at the origin of most of the ores containing metals have caused enrichment of certain structural zones consisting of crystalline or metamorphic rocks, which outcrop only in portions of the Earth's crust that have been strongly uplifted over the course of geological history. Most of the exploitable metal deposits are thus located either in recent mountain ranges such as the Andes Cordillera, the Rocky Mountains or parts of the Alps and Iberian Ranges, in ancient massifs, such as the Mittelgebirge of Central Europe,

1 Loess: a formation of wind-based origin that emerged during the ice ages in the regions at the edge of the ice sheets covering the Northern Hemisphere. On these formations made up of very fine silty particles, which often reach several tens of meters in thickness, very fertile agricultural soils were formed, particularly for cereal crops.

or in ancient shields consolidated and encrusted with various metals, such as Scandinavia and the Canadian, Brazilian, Australian and African shields. In some cases, such as the Lorraine iron basin, metal enrichment comes from sedimentary processes. However, these are rarer situations, representing a small number of metal-bearing deposits in the world.

While some very rich deposits, such as the Rio Tinto copper mines on the Iberian Peninsula, the tin mines of Cornwall in the United Kingdom or the Styrian iron in the eastern Alps, were exploited as early as antiquity, most of the exploitation of metal deposits in the European mountains was developed later, during the Middle Ages, particularly with the demographic and economic boom of the 11th–13th centuries. The exploitation of silver, which is attested from the beginning of the 11th century in the Harz massif and in the Black Forest (Werner and Denert 2004), was to develop in most of the other central European medium mountain massifs (the Mittelgebirge) over the next two centuries. In one of the richest massifs in metals, the Erzgebirge, between Saxony and Bohemia, the first discovery of a vein of silver dates back to 1168, on the site of what was to become the city of Freiberg. Subsequently, the discovery of deposits of silver, and also tin, copper and lead, multiplied and led to the settlement of these uplands which had remained until then entirely forested.

Miners from other mountains that had been exploited for centuries, such as the Harz, flowed into the Ore Mountains and made it the most important mining region in Europe. In the second half of the 15th and early 16th centuries particularly, some of the most important mining towns were discovered: Altenberg (1440), Schneeberg (1471), Annaberg (1496), Marienberg (1521) and St. Joachimsthal (1520, now Jáchymov) in Bohemia. Some of these cities experienced extraordinary growth, comparable to the so-called boom cities of the gold rushes in the American West in the 19th century. A dozen years after its foundation, Annaberg became the most important town in the Ore Mountains with about 8,000 inhabitants (Wagenbreth and Wächtler 2013). Just four years after the discovery of the silver deposit, Joachimsthal (Jáchymov) had over 5,000 inhabitants and by the 1530s it had grown to 18,000 inhabitants, the second largest city in Bohemia after Prague (Hrabanek 2007).

All the other medium mountain massifs of Central Europe have a similar history, marked by the succession of rushes and the foundation of cities that accompany the discovery of new metal deposits. Mining thus successively spread to the Bohemian mountains (Kuttenberg/Kutná Hora), the Slovak Carpathians (Kremnitz/Kremnica, Neusohl/Banská Bystrica, Schemnitz/Banská Štiavnica) and the Eastern Alps (Schwaz in Tyrol and Friesach in Carinthia). Founded by communities from Germany, especially the Saxons, these privileged mining towns were the basis for the spread of Germanic settlement throughout Central Europe.

During the Renaissance, the exploitation of many deposits in Central Europe was made possible by innovations in both water pumping techniques and mineral processing. The development of the cupellation method made it possible to recover silver from copper ores using lead (Lynch 2002). As this process is only profitable if practiced on a large scale, in cities more or less distant from the deposits such as Nuremberg, Villach or Krakow, industrial mineral processing companies (Saigerhütten) were controlled by businessmen such as the Fugger or Thurzo. The latter ended up controlling most of Central Europe's silver and copper production, thanks to the advances they made to princes who, like the Habsburgs, became their debtors (Lynch 2002).

The mining techniques developed in Central Europe were used to exploit the new deposits discovered during the 16th century in various regions of the New World. If El Dorado remains largely a myth, the discovery of many silver deposits, which turned out to be among the richest in the world, nevertheless supported the Spanish conquest of Mexico and Peru. A succession of mining towns were thus founded in the mountains of central and northern Mexico as silver deposits were discovered: the first in Taxco in 1528, then Pachuca and Zacatecas in 1548, Guanajuato in 1558 and San Luis Potosí in 1592 (Avalos-Lozano and Aguilar-Robledo 2017). In the old Inca Empire conquered by Pizarro, the biggest discovery was the deposit of Potosi in 1545, in the Andes Cordillera. Located at 4,800 m above sea level, the Cerro Rico (the rich mountain) would become in a few decades the largest mining city in the world, exceeding 150,000 inhabitants at the beginning of the 17th century.

3.2.2. *Mines, landscapes and the environment in pre-industrial times*

In all the mining regions of Europe and America, the needs of mining not only transformed the population, which was becoming considerably denser, but also put great pressure on natural resources, especially water and wood. In the Ore Mountains, for example, the practice of "Feuersetzen", that is to say, the use of large quantities of wood set on fire in the mine galleries in order to fracture the rock and especially the activity of the very numerous foundries, consumed enormous quantities of wood that needed to be brought from farther and farther away by floating on rivers and streams. Reservoirs were also built to ensure the supply of hydraulic wheels that were used to pump the mine water to keep the increasingly deep mine galleries dry. A whole new landscape appeared, shaped by mining and its consequences (Wagenbreth and Wächtler 2013).

Figure 3.1. *Landscape of the Sierra de Gádor, near Almeria. Old lead mining, in the 1830s, has left visible accumulations of rubble in the background. It destroyed the forest and soils, which explains the desolate character of the current landscape (source: Deshaies (2009))*

Figures from the 16th century give an idea of the importance of wood requirements. For charcoal production in its foundries alone, the city of Freiberg consumed the equivalent of the annual growth volume of more than 30,000 ha of forest in 1 year (Thomasius 1995; Schräber 2008). With such needs, the forests in all the ore mountains ended up being overexploited and reduced to increasingly small areas, making it problematic to supply the mines and smelters. Faced with this question, the Saxon mining administrator

Hans Carl von Carlowitz published the first silvicultural work (*Sylvicultura Oeconomica*) in 1713. In it, he explained how forests can be managed sustainably by harvesting the equivalent of the tree growth each year at most, and for this reason this book is often considered to be the first exponent of the concept of sustainability (Küster 1998; Radkau 2012).

In Mexico, the mountains surrounding the main centers of silver mining still have vegetation reduced to a few shrubs and cacti on slopes that have been emaciated by erosion (Studnicki-Gizbert 2017). The destruction of the wooded vegetation that once covered the slopes of these mountains is largely the consequence of three centuries (1528–1821) of exploitation of mining by the Spaniards. In the Mediterranean regions, more recent examples illustrate the disastrous consequences of mining on vegetation. In the province of Almeria in Spain, whose climate is admittedly particularly dry, the surrounding mountains all present a desolate landscape, almost totally devoid of wood (Figure 3.1). However, this situation is not a product of the climate, but of the exploitation of a lead deposit in the Sierra de Gádor between 1820 and 1880, which led to complete deforestation, not only of the area around the mines, but also of the eastern part of the southern fallout of the Sierra Nevada, the Alpujarra (McNeill 1992). In fact, before the mining boom of the 1830s and 1840s, which made the region the world's leading center of lead production, the Sierra de Gádor was still described by travelers as a mountain largely covered by forests (McNeill 1992). It was the influx of more than 20,000 miners, the need for wood for the galleries, the smelting of ore on site, the supply of firewood, as well as the extension of food crops on the slopes that had remained forested until then, that led to the creation of a large number of forests (McNeill 1992). Soon after it was necessary to import wood from the Balearic Islands and to stop smelting the ore on site, which was exported in its raw state from the 1840s, when mining began to decline, partly due to the lack of wood.

3.3. Mining in the Industrial Age

3.3.1. *The transformations of the industrial energy system*

The increasing pressure on wood resources, which were increasingly limited in Europe in the 18th century, was one of the factors leading to the search for ways to overcome the spatial constraints of the pre-industrial energy system. This system worked on the use of small amounts of flux energy, the Sun, the force of running water and wind and the annual growth of forests and crops. Thus, for the pre-industrial period, we can speak of a system where the quantity of energy used was drawn from available energy and was directly dependent on the supply area, i.e. "energy from space" (Brücher 2009; Smil 2017). Moreover, since the capacity to transport and thus import resources was very limited, the energy required for supply had to be produced from local resources, which severely limited the concentration of settlement. Proto-industrial activities, such as foundries and forges, needed therefore to be dispersed along the watercourses that provided the motive force needed to operate bellows and hammers and, above all, in the middle of forests in order to have sufficient quantities of wood available (Radkau 2012; Deshaies and Baudelle 2013).

With the industrial age, coal mining (Sieferle's "underground forests" (Sieferle 2001)) gave access to a concentrated energy resource, allowing almost total freedom from spatial constraints. Thus, 1 ton of charcoal was equivalent in energy value to about 5 m^3 of wood, i.e. the annual growth volume of 1 hectare of forest in Europe. Contrary to the pre-industrial period, the amount of energy that could be consumed in a given region no longer depended on the surface area available, but on the capacity of society to access fossil energy deposits, whether local or remote. Indeed, the successive technical innovations of the industrial era not only made it possible to use increasingly diversified fossil energy sources (coal, then oil, natural gas and atomic energy), but they also saw a continuous increase in trade and transportation capacity (Deshaies and Baudelle 2013).

As a result, the energy supply was no longer dependent on local production surplus, but on a network linking a region to fossil energy deposits. The resources contained in the deposits were thus exploited (coal mines or hydrocarbon wells) and transformed into energy (e.g. in a thermal power plant), or into fuel (in a refinery), to be then distributed over a territory by a network (power lines or service stations). This is why, unlike the pre-industrial system, it was an "energy for space" production system (Brücher 2009). This system had general economic and geographical repercussions, since it was not only energy resources, but also minerals, industrial and agricultural products that were transported in increasing volumes through increasingly extensive networks around the world. There was thus a growing spatial dissociation between the places where mineral resources were extracted and the places where these materials were consumed. While certain precious metals, particularly gold and silver, were already transported over long distances in the past, more abundant metals, such as copper and especially iron, were then also traded internationally over longer distances and in ever greater volumes. Instead of being melted down on the spot, as was the case in the past, they were now also traded internationally over longer distances and in larger and larger volumes. As a rule in pre-industrial times, iron, the consumption of which was increasing considerably, was transported *en masse* to the coalfields, where a large part of the iron and steel industry settled.

3.3.2. *Birth and development of coalfields*

The boom in coal mining in the last quarter of the 19th century changed the landscape of the coalfields, which, within a few decades, became densely populated industrial regions. Working-class towns developed around the increasingly extensive and productive mining sites. Industries requiring large quantities of coal, both as a raw material and as energy, were concentrated in coalfields and attracted a large workforce from increasingly remote areas.

The Ruhr area in Germany is one of the most spectacular examples of this development. While its development was much later than that of the British Black Country or even Silesia, it was particularly rapid because of its expansion and the growing needs of the German economy at the end of the 19th century. In 1840, coal production of less than 1 million tons left little mark on the Ruhr landscape, which remained deeply rural. Half a century later, in 1890, the boom in mining, which already produced more than 35 million tons of coal and employed 130,000 miners, greatly transformed the landscapes of the southern part of the basin where mining was concentrated. In 1913, on the eve of the World War I, the Ruhr, which produced 114 million tons of coal and employed 444,000 miners, became the largest industrial region in the world and, with more than 3 million inhabitants, one of the largest urban concentrations in Europe.

The landscape was then dominated by slag heaps, headframes, blast furnaces and a multitude of factory chimneys spewing black clouds from coal combustion. The region's environment was thus profoundly transformed, subject to considerable air and water pollution. This evolution was reproduced in most of the coalfields of Europe and North America, which remain densely populated areas to this day, even long after the coal mines have been closed and many of the industries which supported them have also disappeared.

3.3.3. *Conquest and development of new metal deposits*

The economic and demographic boom of the industrial era also led to an increase in the demand for metals, particularly for the six metals that were most commonly exploited at the time, iron, copper, lead, tin, gold and silver, for which demand was exacerbated by both economic growth and technical progress. The demand for copper, for example, was increasing sharply with the development of the use of electricity. The discovery of new deposits of precious metals gave rise to a succession of rushes into new territories, more or less hostile and still almost deserted, or in any case very sparsely populated. The cycle of rushes began with the discovery of gold in the Yenisei basin near Krasnoyarsk in Siberia, almost 20 years before the famous California rush in 1849, followed by the rush in Australia in 1851 and the rush in South Africa in 1873 (Coulson 2012).

The settlement of the mountains of the American West in the second half of the 19th century was mainly due to the gold rushes that brought tens of thousands of adventurers to California (1849), Colorado (1858), Montana (1864), British Columbia (1858) and then Alaska (1902). After the discovery of Californian alluvial gold, which was exhausted within a few years, prospectors went to all the mountains of the American West in search of the fabulous mother vein, i.e. veins containing a high enough concentration of gold or silver to be extracted with the available techniques.

However, the mineral wealth exploited was most often ephemeral, as the technical means of the time did not allow the metal to be recovered when its concentration was too low. Also, after the exhaustion of gold placers or superficial veins, mining towns declined as quickly as they had been created. Many of them became totally abandoned ghost towns. When they did not disappear completely, their population decreased enormously. A famous example is Virginia City in Nevada. Founded in 1859, following the discovery of the famous "Comstock Lode", the first large deposit of American silver, the city had, at its peak in the 1880s, more than 30,000 inhabitants and rivaled San Francisco. The depletion of the mines before the end of the 19th century led to the rapid decline of the city, which had a population of just more than 800 in 2010.

The era of the pioneers was quickly followed by the era of businessmen, such as Andrew Carnegie or the Guggenheims, who took control of the exploitation of many deposits, enhanced by the use of new techniques (Lynch 2002). Indeed, once the higher concentration ores were exhausted, the American West, and also northern Chile, interior Australia and southern Spain, were going to be the first to exploit these deposits to enter a new phase of development, related to the exploitation of low-concentration copper mega deposits. These are mainly copper porphyry deposits, containing hundreds of millions of tons of ore, with grades generally below 2% metal (Neukirchen and Ries 2014).

A whole series of techniques were developed in the last quarter of the 19th century to exploit these deposits, whose economic profitability required the extraction and processing of very large volumes of ore. These included hydro-metallurgical processes, such as leaching and electrolysis, which make it possible to recover almost all of the metal contained in low-grade ore. However, very powerful extraction techniques also had to be developed. Dynamite, developed by Alfred Nobel in 1867, found a privileged field of use in mines. Extraction machines and increasingly large crushers were set up on large deposits. For this reason, at the beginning of the 20th century, open-pit mining began to be developed in various deposits in North America and Chile. Mining became industrialized and stabilized on some sites, giving rise to new landscapes, consisting of large excavations, ore processing infrastructures and, in some cases, nearby smelters.

A number of the large mining deposits that were in operation throughout the 20th century and often to this day were brought into production at that time. These include the Butte[2] copper complex in Montana, operated by Anaconda, and the huge Bingham Canyon mine in Utah, where open-air extraction started in 1906 and still occurs today (Berger 2002). Although production is now surpassed by new mines opened at the end of the 20th century, it is nevertheless one of the largest excavations in the world. Like other Chilean mines, however, it is still mainly a place of extraction, with the partially concentrated metal being transported to the Pacific coast for export to American and European industries. As a result, the settlement of mining deposits located within vast continental spaces, or in mountains with a hostile climate, has mostly been weak and unstable, as they are quite strictly dependent on extraction activities. Although mining was still relatively labor-intensive in the first half of the 20th century, this changed later with the progress of mechanization, particularly the use of increasingly large trucks to transport ore and excavated material. At Bingham Canyon, for example, there are only 2,000 employees today, compared to more than

2 Begun in 1884, copper mining at Butte ended in 1983, leaving a huge open pit mine more than 600 m deep.

15,000 in the 1920s. In addition, these large mines have generally remained mining areas, generating little or no industrial activity on site, apart from initial ore processing. The need for a cheap source of energy to smelt the ores and an abundant workforce meant that industry was unable to set up shop on these deposits, which are often located in the middle of desert regions.

3.4. Contemporary mining transformations and challenges

3.4.1. *Geographic trends in mining*

The second half of the 20th century was a period of profound change in mining that had great geographical repercussions. It was marked by unprecedented growth in the extraction of all the world's mineral resources, as a result of the enormous economic and demographic boom of this period (Smil 2014). The production of the metals most used by the industry, such as iron and copper, tripled or quadrupled between 1950 and 1970 and doubled again between 1970 and 2000 (Deshaies and Baudelle 2013). The beginning of the 21st century, which saw the economic boom in China and various emerging countries, saw an unprecedented acceleration in the exploitation of mineral resources. During the first 15 years of the 21st century, world production of iron and copper, as well as that of many other metals (many of which had been little used until then, such as lithium or rare earths) doubled or tripled (Smil 2014). Coal, which from the mid-1960s was no longer the world's primary source of energy, has also seen its production double since 1980, reaching more than 8 billion tons in 2012. This growth is driven by the place of coal as the primary source of energy for the production of electricity. This is why the decline in the share of coal in the energy mix and the closure of most mines are strictly European phenomena, which do not reflect the general trend.

At the same time, there is indeed a geographical redistribution of former mining operations at the expense of the formerly industrialized countries, where the production of most mines is declining, while new deposits are being discovered or are gaining ground in emerging, developing and even some developed countries. The example of coal is a perfect illustration of this evolution. In 1981, the world's largest coal producers were the USSR, followed by the United States and China, then Germany and Poland (Table 3.1). Three-quarters of production was mined in developed countries, with Europe alone still accounting for 30% of the world total. In 2017, China was by far the world's largest coal producer, supplying almost half of the world's coal. It was followed by India, the United States, Australia, Indonesia and Russia. Seventy percent of global coal production came from the Asia–Pacific region, while Europe's share was just 7%. In most European countries, coal mining has ceased. Only Poland has remained an important coal producer, while in Germany, several Central European countries and Greece, only lignite is mined in large open-cast mines (Deshaies 2007).

Coal production in 1981 (Mt)		Coal production in 2017 (Mt)	
USSR	778	China	3,523
USA	747	India	716
China	621	USA	702
Germany	493	Australia	481
Poland	198	Indonesia	461
South Africa	130.4	Russia	411
India	130	South Africa	252
Australia	128	Germany (lignite)	175
United Kingdom	127	Poland	127
Global production	3,914	Global production	7,727

Table 3.1. *Major coal producers in 1981 and 2017 (source: BP Statistics (2018))*

3.4.2. *Decline and changes in former mining regions*

The evolution of former mining regions can be illustrated through the future of European coalfields. The decline in coal mining since the 1960s has caused a profound crisis in the coalfields, the economy of which, usually highly specialized in mining activities and steel production from coke, has found it difficult to reconvert. Indeed, as long as coal remained the main source of energy and the basis of the iron and steel industry, the coalfields remained economically strong regions with few activities unrelated to mining. Guy Baudelle (1994, 1995) has shown that the mining production system was at the origin of a true spatial system "entirely designed according to the needs of the mine alone". As a result, this economic and spatial system was characterized by its rigidity and inability to evolve from within without undergoing a major break, described as a "creative break" by J.-M. Holz (1987). It was then a question, for the economic and political leaders in place, of deliberately breaking with the practice of supporting traditional industries, or even sacrificing them, in order to favor the development of new activities.

Most often, however, this creative break came only late, when mining activities had already declined sharply, leaving a legacy of wasteland and abandoned mine tiles, long considered unattractive (Figure 3.2). When the new activities started to develop, they could not compensate for the decline of the traditional branches. When local and national politicians realized that the decline in coal production and the number of people employed was inevitable, they tried to reconvert the coalfields by encouraging the development of new industrial activities unrelated to coal. From the end of the 1960s, industrial branches such as the automobile and chemical industries, which were then booming, benefited from national and European financial aid to establish themselves, either in the coalfields themselves, or more widely in nearby regions.

However, what has long been presented as a success story of coalfield conversion is no longer really topical as, since the end of the 1990s, the automotive industry itself has been experiencing difficulties and staff reductions in line with productivity gains. Generally speaking, the re-industrialization of coalfields has only been partially successful. In the long term, the balance of industrial jobs destroyed compared to jobs created still shows a strongly negative balance. This is why the former coalfields share the sad privilege of being areas where unemployment rates are well above the national average, as well as the average for their region: more than 15% in 2018 in the employment areas of the Nord-Pas-de-Calais coalfield. In Lorraine, the unemployment rate in the Forbach employment zone in the coalfield, at over 13%, is the second highest in the region. Finally, in the Cévennes, the employment zone in the former Alès coalfield is characterized by the second highest unemployment rate in France (16.4%). In Germany, the Ruhr area also has unemployment rates well above the national average, although much lower than in France. At the end of 2017, the unemployment rate in the Ruhr reached an average of 9.5% compared to the national average of 5.3% (Deshaies 2019).

Figure 3.2. *Former mining site of the Carreau Wendel in the Lorraine coal basin. The former mine site, which was closed in 1986, became a mine museum in 2006. It contains all the classic elements of a coal mine: headframes, coal washing and slag heap in the background. Only the Schlamms ponds have disappeared (source: Humbert (2009))*

High unemployment or inactivity rates in the vast majority of coalfields go hand in hand with a less skilled workforce. As a result, average income levels in the coalfields are still among the lowest in their region, due to a high proportion of poor people. The lack of a new economic dynamic, which could create enough jobs to replace the losses of the coalfields, has led to a demographic decline in the coalfields (Deshaies 2019). This decline began more or less early and was unevenly significant depending on the chronology of mine closures, the geographical location of the basins and the demographic situation of the country. However, it appears to be fairly generalized, because the coalfields, which have been the centers of immigration since the industrial revolution, are becoming, on the contrary, areas of emigration. Young people, particularly when they have acquired a high level of education, leave the coalfields to find jobs corresponding to their qualifications in the major metropolises. The result is an aging population that ends up strongly eroding the natural balance, which has long remained positive. The end of former mining operations has thus set the coalfields on a long process of demographic decline, and it is not yet known whether it can be halted in the future.

Finally, the coalfields have not really overcome the handicaps deriving from a degraded image, both due to the legacies of mining and the decline of these activities. Although new redevelopment projects have been developed to enhance the industrial and mining heritage and change the negative image of these regions (Gelhar 2010; Wirth *et al.* 2012), this transformation has not yet resulted in what could be called an economic renaissance. While some coalfields, such as the Ruhr and, to a lesser extent, the Nord-Pas-de-Calais basin, have managed to revalue their mining landscapes and obtain world heritage recognition (UNESCO classification), this has not been enough to generate another form of economy, dynamic enough to reduce the social problems of these former mining regions.

3.4.3. *Extraction boom and risks in new mining regions*

Although large-scale open-pit mining began to develop at the beginning of the 20th century, the number of large open-pit mines increased from the 1980s onward for both coal and metal extraction. A growing share of the United States' coal is extracted from deposits in Wyoming, in huge open-cast mines. Indonesia and Australia have become the world's leading coal exporters, with major discoveries in East Kalimantan and Queensland, or New South Wales, respectively, since the late 1990s (Figure 3.3). These mines, connected to mineral terminals by rail, are generally located in sparsely populated areas and attract few people, as they remain highly mechanized mining sites, with production exported directly to the major coal consumers China, Japan, India and South Korea.

The low density of the population allows the digging of mines that have considerable extension, sometimes stretching over more than 10 km in length. In Indonesia, they cause a disruption of the landscape, contributing to the destruction of the tropical forest of the island of Kalimantan. The pollution of water by the excavated material and the negative impact on local agricultural production are denounced by environmental NGOs, without this managing to call into question the practices of an economic sector considered essential by the state. In Australia, as much as the impact of the mines themselves, it is also the consequences of the development of tank terminals and the effects of transport on marine environments, particularly those of the Great Barrier Reef, that are denounced by the NGOs. One of the most controversial projects at present is the Carmichael mine in Queensland. The Indian company Adani has proposed to open an immense open-cast mine that would be one of the largest in the world, from which 60 million tons of coal could be extracted per year for 60 years. The NGOs campaigning against this mine not only contest the local environmental damage it will cause, but also the relevance of a project aimed at extracting huge quantities of coal, which contribute significantly to global warming. Despite this controversy, in June 2019, the Australian government approved Adani's project to operate the Carmichael coal mine. This approval could pave the way for new mining projects in the hitherto undeveloped Galilee coalfield (Figure 3.3).

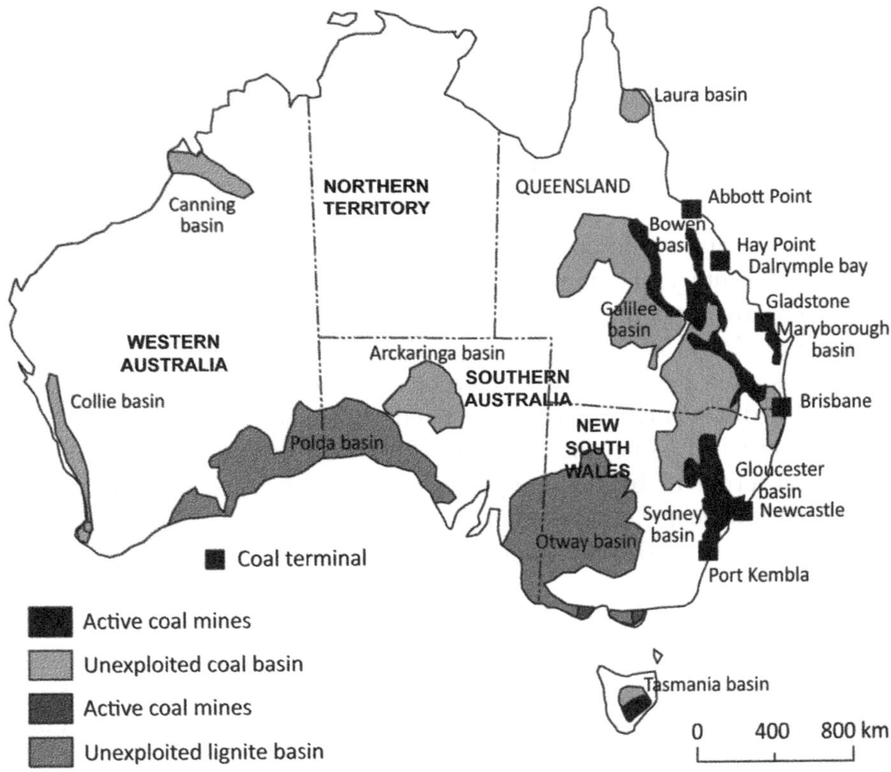

Figure 3.3. *Coal basins and terminals in Australia. For a color version of this figure, see www.iste.co.uk/fizaine/mineral2.zip*

The surge in demand has led to intensified mining on deposits that have been known and exploited for a long time, as illustrated by the example of the Chuquicamata copper mine (Figures 3.4 and 3.5) in Chile. Over the last 30 years, several new mines have been discovered there and the considerable volumes of waste material generated by this extension of exploitation are burying the mining town, whose inhabitants have been transferred to the nearby town of Calama, Chile. In spite of these considerable investments to intensify the exploitation of the deposit, the mine's production is in decline and is expected to continue underground from 2019. After having been the most productive copper mine for a long time, it is now only the fifth mine in Chile, far behind Escondida, a mine opened in 1988 that is now the largest copper mine in the world.

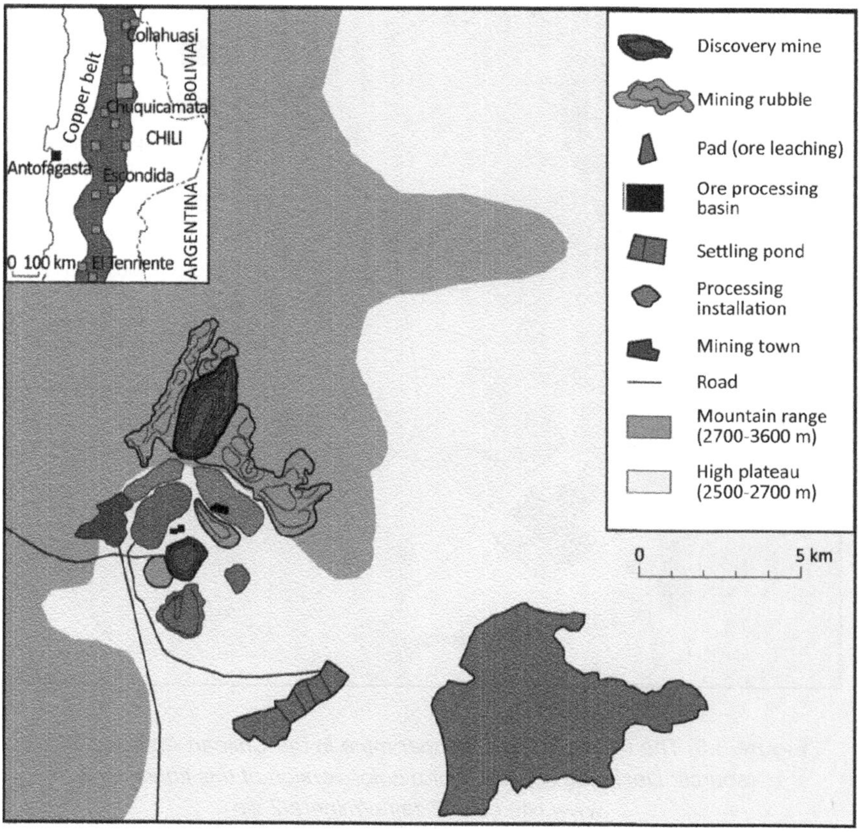

Figure 3.4. *The Chuquicamata copper mine in the Chilean Andes, 1988 (source: Deshaies (2019)). For a color version of this figure, see www.iste.co.uk/fizaine/mineral2.zip*

COMMENT ON FIGURE 3.4. – *Opened in 1913, the discovery did not cease becoming deeper over the course of time, becoming one of the biggest excavations dug by humans. The mine was then at the peak of its production, which exceeded 1 million tons of pure copper per year. One can notice the extension of the accumulation of cuttings, as well as the vastness of the tailings impoundments for the ore processing residues.*

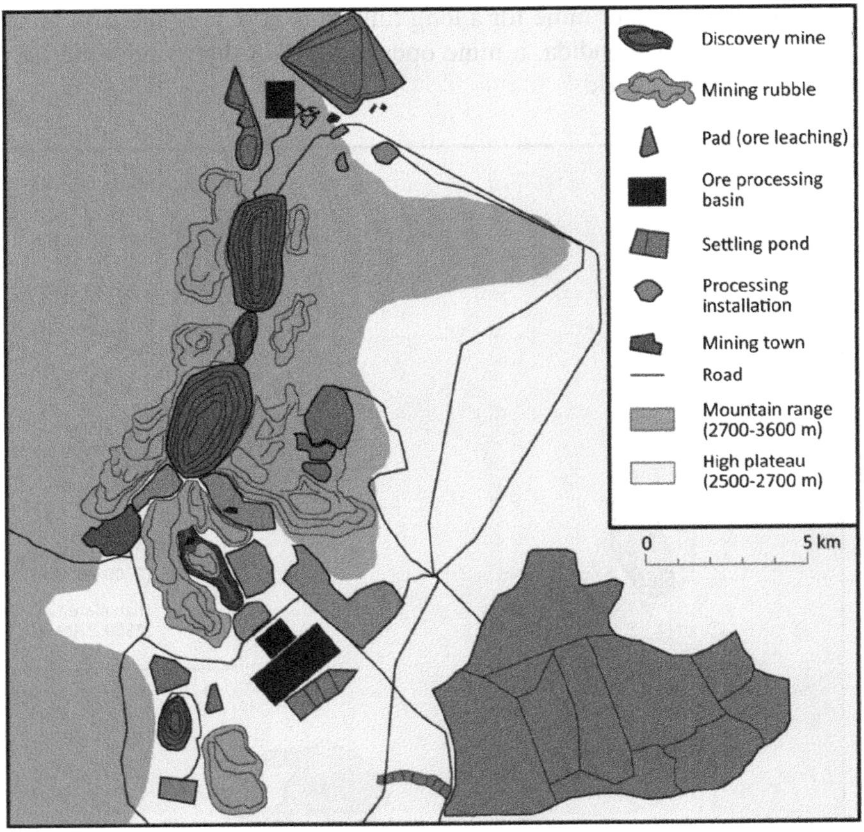

Figure 3.5. *The Chuquicamata copper mine in the Chilean Andes in 2018 (source: Deshaies (2019)). For a color version of this figure, see www.iste.co.uk/fizaine/mineral2.zip*

COMMENT ON FIGURE 3.5. – *Thirty years later, mining operations have expanded significantly, with two new open pits, one of which, north of the original site, is very large. This is the Radomiro Tomic mine. This has enabled annual production to be maintained at more than 800,000 tons of pure copper per year for both mines combined.*

Since the 1990s in particular, the depletion of high-grade metal deposits has led to the exploitation of gold, silver and copper deposits in very low-grade ores, such as in Nevada, Peru and Argentina (Jebrak 2015). The sharp rise in gold and silver prices during the 1970s and even more so since the beginning of the 21st century has ensured the profitability of the considerable investments required to dig large mines and process huge volumes of ore by cyanidation on a large scale in order to recover limited quantities of precious metals. Today's largest gold mines, for example, exploit ores with average grades of no more than 1 or 2 grams per ton. However, the extraction capacity of these mines can reach millions of tons of ore per year, resulting in a production of up to several tens of tons of pure gold (Deshaies 2018).

The increase in global demand and metal prices has led to the multiplication of mining projects, not only in areas where long-established known deposits are located, but also in new areas where low-grade deposits have been discovered that were previously uneconomic to mine. While some of these deposits are in desert regions, such as northern Chile, where mining activity does not compete, some mining projects have developed in more densely populated areas, where they potentially threaten agricultural activities. Fears of dramatic pollution of waterways by effluents from mines are far from unjustified, since they have materialized in the past few years. The pollution of a tributary of the Guadalquivir in 1997, following the accidental discharge from a tailings impoundment at the Aznalcollar mine in Andalusia, is an illustration of these risks. A few years later, in 2000, a similar incident occurred at the Baia Mare mine in Romania, causing dramatic cyanide pollution of the Tisza, a tributary of the Danube. Among the most catastrophic recent examples are the repeated breakdowns, in the space of three years, of two tailings impoundments in the Brazilian iron mines of Minas Gerais. The accident that occurred at the Samarco mine in November 2015, which killed some 20 people and seriously polluted the Rio Doce over 650 km, was repeated in a much more dramatic way at the Bento Rodrigues mine in January 2019, when the tailings dam burst, killing more than 300 people.

This is why the mining boom since the 1990s has led to an increase in conflicts over projects considered destructive of their living environment by local populations (Blaikie 1995). These conflicts are particularly numerous and violent in Latin America, where many mining projects have been launched by foreign companies, most often from Canada, the United States and more recently Canada (McMahon and Remy 2001). This has contributed to the exacerbation of conflicts, as foreign companies are accused by opponents of mining and extracting wealth that does not benefit local communities, whose agricultural environment and economy are at risk of being destroyed by pollution from mining. For this reason, this boom in mining projects in Latin America in particular is often condemned and described as "extractivism", by transferring a term used to qualify the excessive exploitation of biosphere resources (especially in the Amazon), for which the term was originally used (Svampa 2011; Bednik 2016).

The Observatory of Mining Conflicts in Latin America, which brings together a number of these NGOs, identified 280 conflicts (up from 217 in 2014) in 2018 in 19 countries. Chile (47), Mexico (45) and Peru (42) alone account for almost half of these conflicts. This high level of conflict is a direct consequence of the increase in investment in the territories of indigenous communities in these countries. Peru is an emblematic example. The strong growth in mining production in Peru, which has once again become an Eldorado, and in particular the world's leading silver producer and sixth largest gold producer, is linked to the development of new low-grade deposits in the form of very large open-pit mines, such as the Yanacocha mine (Figure 3.6), near Cajamarca in the north of the country. In Peru, nearly 90% of the area under mining concession has been mined since 1990 and most after 2000 (Bebbington *et al.* 2008, Bebbington and Williams 2019).

Conflicts around mining or hydrocarbon extraction projects are often extremely violent and have frequently led to clashes in which as many as 53 people have been killed in Peru since 2011. There have been conflicts over the expansion of several gold or copper mines in Peru (Yanacocha, Lagunas Norte, Tía María, Las Bambas), and the proposed Tambo Grande gold mining project filed in 1999 by the Canadian company Manhattan Minerals was finally abandoned. It had faced strong opposition from the local populations who feared that the planned mine would cause water pollution and a partial dewatering of the Piura River on which their irrigated agriculture depended. The opening of the Las Bambas copper mine in December 2015, operated by the Chinese company MMG, was preceded by violent demonstrations, resulting in the death of four opponents who wanted to prevent the construction of an ore processing plant, which was considered a threat to agricultural water resources.

Figure 3.6. *The Yanacocha Gold Mine near Cajamarca, Peru (source: Deshaies (2009))*

COMMENT ON FIGURE 3.6.– *This large open pit mine, in operation since 1992, is representative of the new generation of mines that have developed in new areas with no mining tradition. Located at an altitude of more than 4,000 m, it has generated strong protests from farmers fearing pollution of water resources.*

This radical opposition to the risks of water pollution from mines is also based on a more fundamental criticism of the consequences of the over-dependence of these countries' economic development on middling resources. Indeed, many developing countries with significant mineral or energy resources have experienced less economic development than countries without subsoil resources (Gylfason, 2001; Davis and Tilton, 2002). This "resource curse" thesis has been much argued by some economists (e.g. Auty 1993, 2001; Sachs and Warner 1995, 1997, 1999, 2001) who have tried to show the mechanisms by which the exploitation of abundant mineral resources can have a negative influence on economic development. Even if this "resource curse" is not necessarily inevitable (Wright 1990; Findlay and Lundahl 1999; Wright and Czelusta 2004; Jebrak 2015), this theory is also widely used by NGOs and the populations they support in their fight against mining projects. More recently, this criticism of the harmful consequences of the exploitation of mineral resources on development has been reinforced by arguments about the need to restrict the consumption of resources in order to limit the impact of humans on the planet. For activists opposed to mining, it destroys the environment without allowing for socio-economic development, the overall negative consequences of which are disputed. Consequently, it is better to ban it and leave mineral resources in the ground.

Some conflicts between local communities and mining companies have received significant media coverage, and environmental NGOs have developed campaigns to draw public attention to the environmental and social nuisances of major mining projects. Gold mining is particularly targeted and, through various media campaigns, NGOs are trying to make consumers and manufacturers of objects using gold responsible by asking them to commit to not using gold from mining operations that have a negative impact on local populations and the environment. In recent years, there has been a growing number of initiatives aimed at establishing certifications attesting to the origin of gold from mining operations that are qualified as environmentally and socially responsible.

3.4.4. *The limits of "responsible" mining*

The multiplication of conflicts around mining projects at the beginning of the 21st century has put mining companies under increasing pressure, not only in the countries where the minerals are extracted, but also in those to which they are exported. In 2002, the United Nations Environment Programme (UNEP) thus insisted on the need for mining companies to take into account environmental management and to open dialog with the communities of inhabitants directly concerned by their projects. In 2003, the major international banks defined the "Equator Principles", which make the financing of mining projects subject to compliance with a certain number of social and environmental criteria. In 2008, the International Finance Corporation (IFC), the UN and the OECD proposed that mining companies comply with a number of recommendations on environmental issues and community relations (Jebrak 2015).

In the face of these pressures, mining companies have gradually developed a strategy to show that they have integrated the social and environmental concerns that have been expressed over the past 20 years (Spitz and Trudinger 2009). The objective of this strategy is to obtain a "*social* license to operate"; in other words, to have an authorization to operate granted by the community directly concerned (Thomson and Boutilier 2011; Jebrak 2015). The success of mining projects increasingly depends on obtaining acceptance of this activity by the affected communities. The strategy of major mining companies is to avoid mining conflicts that can not only tarnish their image, but also cost them dearly if the project has to be cancelled. Many companies have therefore committed to a process that aims to open dialog with communities during the exploration phase in order to respond to their concerns and expectations in advance. They have thus equipped themselves with personnel dedicated to communication with the communities, in order to, first, obtain legitimacy, and second, establish bonds of trust (Thomson and Boutilier 2011). Taking into account the expectations of the community, even during the construction of the mine, is a complex process, insofar as the project itself can evolve significantly depending on the results of exploration. Hence, the need for a long-term dialog, as the impacts of the future mine cannot be specified from the outset. Obtaining the social license to operate presupposes that the mining project can more or less completely satisfy the three pillars of sustainable development: controlling the environmental impact, promoting the economic development of the local community and providing social benefits that outweigh the nuisances it generates.

The evaluation of the various impacts of a mining project is, however, complicated and can only with difficulty lead to a consensus between those who place a sacred value on the environment and those who consider that controlled exploitation is possible and even necessary for the development of communities. In the face of this dilemma, mining companies strive to show, through concrete achievements, that their activities can bring material benefits to the communities concerned. These are projects that allow the mining company to demonstrate its environmental responsibility in the mining techniques employed, its social responsibility through actions carried out in favor of the communities and its commitment to environmental restoration or improvement (Spitz and Trudinger 2009).

The environmental responsibility of the companies consists of using the most modern and least aggressive techniques for the environment. In some cases, the new mining project is presented as likely to contribute to the improvement of the environment, particularly water quality (Spitz and Trudinger 2009). In order to develop its Pueblo Viejo project in the Dominican Republic, Barrick Gold signed an agreement in November 2009 with the Dominican government in which it agreed to pay the costs of cleaning up the former mining sites within the scope of its concession.

Social responsibility is exemplified by the support of projects for the marketing of local products that increase the financial resources of the communities. The mining companies have also launched cooperation and community assistance projects, building schools and financing education and care programs, thus providing public services that some states cannot provide. The objective is to build a relationship of trust with local populations. This assistance can also take the form of infrastructure work to develop the local economy. This is the case, for example, in Yanacocha where, since 2006, the former San José open-pit mine has been transformed into a water storage reservoir to supply irrigation canals used by several thousand local farmers. This participation of the mining company in the development of irrigated agriculture in a region with a long dry season allows the mining company to counter the arguments of the opponents of the mine who accuse it of polluting local water resources.

Mining companies are also trying to build a "green" image by financing projects to protect or restore nature, reforest mining areas, rehabilitate old mines or invest in renewable energy. As part of the promotion of the Pascua-Lama project, Barrick Gold has installed a large wind turbine to produce part of the electricity from its Veladero mine at an altitude of more than 4,000 m. It also financed a large wind farm at La Higuera in the Coquimbo region. However, while these projects are presented as intended to provide part of the mine's electricity supply, they are mainly seen as communication tools to illustrate the company's environmental responsibility.

In fact, despite real efforts to limit their impact on the environment and to obtain or maintain acceptance of their project by the communities in the territories where they operate, mining companies are faced with the problems of the scale of current mining operations. Large-scale mining leaves such large legacies in landscapes that their rehabilitation appears to be extremely problematic.

3.5. Conclusion

Mining has played a decisive role in the settlement of areas that are not conducive to agricultural occupation. Many mountainous regions in Europe as well as in America only experienced a spread of settlement following the discovery of metal deposits. In the industrial age, the conditions for the spread of settlement changed radically. Only the mining deposits that served as a support for the development of industry saw their populations become much denser. This was mainly the case for coalfields, as well as a small number of metal deposits located nearby or connected to large urban areas. The mining deposits located in the desert regions of America, Australia or Russia, far from the large population centers, remained essentially pioneer areas, sparsely populated.

The considerable expansion of mining in the second half of the 20th century intensified the production of deposits with the most favorable geological and economic conditions. Wherever technically possible, large-scale open-pit mining was developed, which made it possible to considerably increase productivity and lower costs. From the 1990s in particular, new deposits, located in areas with no mining tradition, were brought into production. Large-scale mining projects have multiplied, particularly in Latin America, Asia, Africa and Australia. The production boom in these new mining regions has largely contributed to the decline of European mines, most of which were closed down in the last quarter of the 20th century.

The irruption of large-scale mining in new, mainly agricultural areas has led to an increase in environmental conflicts, as farming communities fear that the mines will cause dramatic pollution of their water resources. In order to limit the extent of the conflict, the mining companies have developed a strategy to enable them to establish relationships of trust with the local communities. This strategy consists of ensuring their territorial anchoring by making local investments that contribute to improving the living environment of the populations and promoting their economic development.

In fact, the question of the future of territories and populations after the cessation of mining operations is probably the thorniest problem, although little mentioned by mining companies. Even if the pollution can be completely controlled during the period of mining activity, the transformations of the landscapes and more broadly of the living environment of the populations are so important that the resilience of the territory is profoundly altered. After the closure of a mine, which takes place after about 10 or 20 years, depending on the case, the question of the reconversion of these territories into new activities will arise.

3.6. References

Auty, R. (1993). *Sustaining Development in Mineral Economies: The Resource Curse Thesis*. Routledge, London.

Auty, R. (2001). *Resource Abundance and Economic Development*. Oxford University Press, Oxford.

Avalos-Lozano, A. and Aguilar-Robledo, M. (2017). Reconstructing the environmental history of colonial mining. In *Mining North America*, McNeill, J.R, Vrtis, G. (eds). University of California Press, Berkeley.

Baudelle, G. (1994). Le Bassin minier du Nord-Pas-de-Calais après le charbon : la difficile gestion de l'héritage spatial. *Hommes et Terres du Nord*, 1, 3–12.

Baudelle, G. (1995). Le système spatial de la mine, l'exemple du bassin houiller du Nord Pas de Calais. PhD Thesis, University of Lille III.

Bebbington, A. and Williams, M. (2019). Water and mining conflicts in Peru. [Online]. Available at: http://snobear.colorado.edu/Markw/Research/08_peru.pdf [Accessed 30 May 2019].

Bebbington, A., Hinojosa, L., Humphreys, D., Bebbington, D.H., Burneo, M.L., Warnaars, X. (2008). Contention and Ambiguity: Mining and the Possibilities of Development. Working document 57, Brooks World Poverty Institute, Manchester.

Bednik, A. (2016). *Extractivisme. Exploitation industrielle de la nature : logiques, conséquences, résistances*. Le Passager Clandestin, Paris.

Berger, A. (2002). *Reclaiming the American West*. Princeton Architectural Press, New York.

Blaikie, P. (1995). Understanding environmental issues. In *People and Environment*, Morse, S., Stocking, M. (eds). University of East Anglia, Norwich [Online]. Available at: https://www.bp.com/en/global/corporate/energy-economics/statistical-review-of-world-energy.html.

Brücher, W. (2009). *Energiegeographie*. Gebrüder Borntraeger Verlag, Berlin.

Coulson, M. (2012). *The History of Mining*. Harriman House Ltd, Petersfield.

Davis, G. and Tilton, J. (2002). *Should Developing Countries Renounce Mining? A Perspective on the Debate*. Colorado School of Mines, Denver.

Deshaies, M. (2007). *Les Territoires miniers, exploitation et reconquête*. Ellipses, Paris.

Deshaies, M. (2018). Grands projets d'exploitation minière et stratégie des firmes pour se rendre environnementalement acceptables [Online]. Available at: https://espacepolitique.revues.org/2113 [Accessed 11 December 2018].

Deshaies, M. (2019). Les Bassins houillers européens : déclin, problèmes et potentiels de régions post-minières. In *Un après-mine imprévu. Entre craintes et attentes, quand l'exploitation du sous-sol redevient d'actualité en Lorraine*, Raggi, P. (ed.). Presses universitaires de Nancy, Nancy.

Deshaies, M. and Baudelle, G. (2013). *Ressources naturelles et peuplement*. Ellipses, Paris.

Findlay, R. and Lundahl, M. (1999). Resource-led growth – A long-term perspective. The Relevance of the 1870–1914 Experience for Today's Developing Economies. Working document 162, UNU/WIDER.

Gelhar, M. (2010). Altindustrieregionen zwischen Verfall und Neuorientierung. *Geographische Rundschau*, February, 2, 4–9.

Gylfason, T. (2001). Natural resources, education, and economic development. *European Economic Review*, 45, 847–859.

Holz, J.M. (1987). Les vieux bassins industriels : la dialectique ternaire-sclérose-rupture créatrice-renouveau. *112ème Congrès national des Sociétés savantes*. Lyon, 119–127.

Hrabanek, J. (2007). Die Geschichte des Bergbaus im böhmischen Erzgebirge [Online]. Available at: http://www-user.tu-chemnitz.de/~fna/agricola16.pdf.

Jebrak, M. (2015). *Quels métaux pour demain ? Les enjeux des ressources minérales*. Dunod, Malakoff.

Küster, H.J. (1998). *Geschichte des Waldes*. C.H. Beck, Munich.

Lynch, M. (2002). *Mining in World History*. Reaktion Books, London.

McMahon, G. and Remy, F. (2001). *Large Mines and the Community: Socioeconomic and Environmental Effects in Latin America, Canada, and Spain*. World Bank, IDRC, Toronto.

McNeill, J.R. (1992). *The Mountains of the Mediterranean World*. Cambridge University Press, Cambridge.

Neukirchen, F. and Ries, G. (2014). *Die Welt der Rohstoffe*. Springer Spektrum, Wiesbaden.

Radkau, J. (2012). *Holz, wie ein Naturstoff Geschichte schreibt*. Oekom Verlag, Munich.

Sachs, J. and Warner, A. (1995). Natural Resource Abundance and Economic Growth. Working document 5398, NBER.

Sachs, J. and Warner, A. (1997). Natural Resource Abundance and Economic Growth. Working paper, HIID, November.

Sachs, J. and Warner, A. (1999). The big push, natural resource booms and growth. *Journal of Development Economics*, 59(1), 43–76.

Sachs, J. and Warner, A. (2001). Natural resources and economic development: The curse of natural resources. *European Economic Review*, 45, 827–838.

Schräber, D. (2008). *Das Landschaftsbild um die Bergstadt Freiberg zwischen dem 16. Und 19. Jahrhundert, Die Entstehung einer Kulturlandschaft im sächsischen Erzgebirge*. Tk-reprint GbR, Freiberg.

Sieferle, R.P. (2001). *The Subterranean Forest*. The White Horse Press, Winwick.

Smil, V. (2014). *Making the Modern World*. John Wiley & Sons, Hoboken.

Smil, V. (2017). *Energy Transitions*, 2nd edition. Praeger, Santa Barbara.

Spitz, K. and Trudinger, J. (2009). *Mining and the Environment*. CRC Press, Boca Raton.

Studnicki-Gizbert, D. (2017). Exhausting the Sierra Madre. In *Mining North America*, McNeill, J.R., Vrtis, G. (eds). University of California Press, Berkeley.

Svampa, M. (2011). Néo-développementisme extractiviste, gouvernements et mouvements sociaux en Amérique latine. In *Problèmes d'Amérique latine*, volume 3. Choiseul, Paris.

Thomasius, H. (1995). Der Einfluss des Bergbaus auf Wald- und Forstwirtschaft im sächsischen Erzgebirge bis zum Beginn des 19. Jahrhunderts. Sächsischer Forstverein, 17.

Thomson, I. and Boutilier, R. (2011). Social license to operate. In *SME Mining Engineering Handbook*, Darling, P. (ed.). Society for Mining, Metallurgy and Exploration, Littleton.

Wagenbreth, O. and Wächtler, E. (2013). *Bergbau im Erzgebirge*. Springer Spektrum, Wiesbaden.

Werner, W. and Dennert, V. (2004). *Lagerstätten und Bergbau im Schwarzwald*, Lande-samt für Geologie. Rohstoffe und Bergbau, Baden-Württemberg.

Wirth, P., Cernic, M., Fischer, W. (2012). *Post-Mining Regions in Central Europe, Problems, Potentials, Possibilities*. Oekom Verlag, Munich.

Wright, G. (1990). The origins of American industrial success, 1879–1940. *The American Economic Review*, 80(4), 651–678.

Wright, G. and Czelusta, J. (2004). The myth of the resource curse. *Challenge*, 47(2), 6–38.

PART 2

Action Levers

4

Maintaining or Even Developing the Mining of Mineral Resources in Europe: The Case of Wallonia (Belgium)

Johan YANS[1,2]
[1] *Institut Transitions, Lyon, France*
[2] *ILEE, University of Namur, Belgium*

4.1. Introduction

Several chapters in this book demonstrate the omnipresence of geological resources in our daily lives. With our current state of technology and way of life, *Homo sapiens* "need" to extract abundant quantities of mineral materials. Admittedly, this claim remains debatable: in this transition period, reducing our needs, increasing recycling, finding "technological" alternatives, drawing up and legislating are all ways to limit the extraction of geological resources, which are by definition non-renewable (see Chapter 5). In fact, the mining of these resources continues to intensify, with rapid population growth and changes in consumer lifestyles: the deposits of the future will be increasingly difficult to exploit (Mudd 2007). Over several centuries, *H. sapiens* have logically already exploited all or part of the most easily accessible deposits.

Today, a large number of the mineral materials that feed European industries come from subsoils outside Europe. The European citizen is therefore "free" of the nuisances caused by extractions, while "benefiting" from products made from geological resources. On the other hand,

Mineral Resources Economics 2,
coordinated by Florian FIZAINE and Xavier GALIÈGUE.

European industries in particular are highly dependent on foreign supplies. The purpose of this chapter is to analyze the arguments considered to justify maintaining, or even strengthening, extractive activity in Europe, based on the case of the Walloon region (Belgium). This issue is also relevant in France, probably in a manner similar to that described in Wallonia, even though subsoil management is a strictly regional competence in Belgium[1], whereas this competence is very centralized in France.

4.2. Geological resources in Wallonia

Wallonia is a mining region of Belgium: for thousands of years, many of its mineral materials have been extracted (flint, gold, fluorite, "marbles", phosphates, iron, lead, zinc, baryte, coal, clays, etc.). Geology developed there in the 19th century, in close connection with the industrial revolution based in particular on coal and metals. Several Walloon localities have given their names to geological periods recognized internationally, even today, by the International Union of Geological Sciences (IUGS): Dinant (Dinantian), Viséen (Viséan), Tournai (Tournaisian), Frasnes-les-Couvin (Frasnian), etc. Several tens of millions of tons of mineral materials are still extracted in Wallonia, but no more metal.

4.2.1. *Extraction of mineral materials other than metals*

Limestones, dolomites, chalks, sands, clays, sandstones, shales and "porphyry" are exploited on about 150 sites listed as extraction zones in the sector plan of the Walloon Region (Yans and Dekoninck 2016; Goemaere *et al.* 2020). The materials extracted in this way supply, regionally or abroad, a wide variety of sectors such as ornamental stone, the steel industry, cement, pharmaceuticals, glassware, concrete, water and gas treatment, agriculture and sugar refineries. The applications are numerous but still not well-known to the public. Several world leaders in this field are Belgian (e.g. Lhoist and Carmeuse). However, the economic balance sheet at the Belgian level shows that imports of mineral materials (including qualities extracted in Belgium) are higher than exports. Like European industry, Belgian industry is dependent on geological materials (including non-metallic materials) extracted outside its territory.

1 In Belgium, the three regions (Flanders, Brussels-Capital and Wallonia) manage their subsoil, including geological resources.

4.2.2. Metal extraction: a problem on several spatiotemporal scales

There are no more quarries/mines exploiting metals in Wallonia. The last iron mine in the Meuse valley (Couthuin) closed in 1946 (Denayer *et al.* 2011) and the last quarry for a metallic substance dates back to the end of the 20th century (baryte extraction in Fleurus). Historically, Wallonia was a major producer of zinc during the 19th century and the first half of the 20th century (Dejonghe 1998). During the first half of the last century, mining activities gradually faded away. Belgium, and in general several European countries, then "exported" the extractions to the colonies: the Belgian Congo supplied – and still supplies (Congo, Zaire or the Democratic Republic of Congo, according to the period) – Belgium, Europe, and even the world for certain metals.

From then on, the Belgian citizen considered the problem of geological extraction as:

– (pseudo/post-)colonial;

– associated with distant regions/populations;

– related to a country's political systems, undoubtedly corrupted and with networks not very scrutinized.

Public awareness of the necessities and consequences of mining extraction was then gradually diluted into other emerging issues.

However, several geological resources remain very present in Wallonia, France and Europe. European countries have thus preserved and developed significant mining activities (e.g. Sweden, Finland, Poland, Spain, Ireland and Bulgaria): there are several dozen mines exploiting metallic ores in Europe. These mines can only supply a low proportion of the needs of European industries: most of the supply is provided by mines/quarries located outside Europe.

In the current and future context of increasing demand for geological substances, several voices plead for the "return" of metal extractions in Wallonia (Goossens 2014) and for the intensification of the mining of non-metallic raw materials, including fossil fuels such as coalbed methane (Yans *et al.* 2013), which will not be dealt with in this book. A similar reflection has been envisaged in France (Charles *et al.* 2017).

4.3. Extension of sites/quantity of mining?

Whether it is to maintain/develop the supply of non-metallic materials or to (re)launch metal extraction, Wallonia, like Europe, would be led to intensify the mining of its subsoil. With this in mind, mining companies regularly submit requests to the Walloon Region's administration to modify (or "extend" in the case of sites already in operation) the sector plan for the extraction zone for non-metallic materials. These requests are subject to a detailed impact study and examination by the local/regional authorities. In addition, over the past 15 years or so, mining companies have been interested (again) in metal resources/reserves in Europe. Wallonia is no exception: in 2017, a permit for exploration (please read "exploration" and not "extraction") was requested by the Walzinc company for the famous sites of La Calamine[2], Plombières (the name of the village says a lot about its subsoil; "plomb" means "lead" in French), Hombourg and Gemmenich. The request concerned a permit to search for non-ferrous metal ores (zinc-lead and also the by-products indium, germanium and gallium) by geophysical and geochemical prospecting and drilling.

The main arguments put forward, generally speaking, in favor of such "new" (re)exploitation of metals in Europe (particularly in Wallonia) are described below.

4.3.1. *Exploit existing and well-characterized metal resources/reserves*

For about 10 years now, the European Union and the Member States have been drawing up an inventory of geological raw material resources/reserves likely to be (re)exploited on the territory of metropolitan Europe. Several scientific projects have thus been undertaken by national geological services, universities and mining companies. The resources exist, which has led several scholars to consider the problem of critical mineral materials in Europe not as a shortage (*the resource is difficult to access, or even absent*) but as an incursion (*we do not give ourselves the means to exploit the resource*). The possibility of new prospecting/extraction on the Walloon sites of

2 The La Calamine mining site, discovered in 1829 and mined until about 1879, is well-known in the mineral extraction world: the term "La Calamine" characterizes a zinc deposit of meteoric origin, i.e. generated by interactions between the rock and meteoric fluids. *Calamine-type ore* is therefore a term widely used, worldwide, in the mining sector.

La Calamine, Plombières, Hombourg and Gemmenich (see above) is based on fairly solid scientific bases; the ore is relatively well-characterized thanks to:

– knowledge of previous extractions;

– surveys already carried out on these sites for several decades (see the summary in Dejonghe (1998));

– numerous petrographic, mineralogical, geochemical and associated geological studies (see the synthesis in Coppola *et al.* (2008)), including for strategic metals (Goffin *et al.* 2015; Choulet *et al.* 2019).

However, some argue that the resource would be less easy to exploit in Wallonia than in other places where the orebodies have never been mined in the past: the Walloon sites have already been the subject of historical mining (about 1,100,000 tons of zinc and 130,000 tons of lead; Laloux *et al.* 2000). The deposits were mined to depths of between 100 and 230 meters before being abandoned. The reasons for these abandonments are the narrowing of the veins at depth and their depletion of lead and zinc in favor of the less interesting pyrite (FeS_2) and significant water inflows requiring the digging of dewatering galleries and dewatering to the surface (Ruthy and Dassargue 2009). On the other hand, others mention that very interesting orebodies remain accessible, especially with modern techniques: Goossens (2014) reported 1.7 million tons of ore with grades of 11% zinc, 2% lead (European mine averages are around 6.7% zinc and 1.7% lead) and 30 g/ton silver.

In mainland France, the geological substances with the highest probability of mining development are tungsten, antimony, fluorite and gold; the ores would be intact (i.e. not previously mined) at a depth of 300 meters (Vic 2017).

4.3.2. *Promoting a short circuit*

The transport of mineral raw materials (tens/hundreds of millions of tons depending on the type of material, especially for non-metallic materials) generates financial and ecological costs. Exploiting the minerals in Europe, to process them in a short circuit in Europe, is a responsible approach that could limit the impacts related to transport and encourage the circular economy. Moreover, the extractive activities of some "developing" countries are more polluting than the modern techniques that would be used in Europe (and which are already used for imported minerals). Furthermore, the impact studies envisaged in Europe are not commensurate with those (if any) in several "distant" countries. Sustainable development is not a local but a global consideration. In all ecological rigor, extraction should be considered in the "least polluting possible" way, rather like processing plants and markets. However, if the citizen is ready to consider, to varying degrees, a short circuit for food products, he or she is not very comfortable with short circuits for mineral resources.

4.3.3. *Promoting alternatives to the sometimes deplorable extraction conditions in some regions of the world*

It is clear that any future metal extraction in Europe would be carried out under much better working conditions than those currently applied in some parts of the world, especially in so-called "artisanal" activities. However, should we not favor/impose decent mining conditions in these remote regions, rather than advocating for mining in Europe? Several NGOs are tirelessly campaigning in this direction.

4.3.4. *Stimulating the local economy/employment*

In Wallonia, the extractive industry generates approximately 2,500 direct jobs and 16,000 indirect jobs (Poskin 2010). An important characteristic of the sector is that the jobs are diversely qualified, from highly qualified executives to unqualified personnel. New extraction sites will generate employment. However, some voices predict that the jobs created will be limited in number and will not compensate for the loss of jobs in sectors directly affected by the return of extraction sites, such as tourism. For example, the region of La Calamine has strongly developed the tourism sector in recent decades, which would undoubtedly be affected by mining activities (landscape impact, cartage, reputation, etc.), according to intensities that have yet to be evaluated.

4.3.5. *(Re)discovering a degree of supply independence for the industry*

The European industry is aware of its dependence on critical/strategic geological materials, most of which are extracted outside Europe. (Re)exploiting in Europe, including in Wallonia, matters that are dealt with in Europe seems a judicious approach, even if increasing globalization tends to limit this argument. Over the past 15 years or so, awareness of this aspect has generated numerous articles, conferences and reports in Europe (European Commission 2018), including in France (De Marsily and Tardieu 2018). A similar approach is being taken in the United States (Nassar *et al.* 2020).

4.3.6. *Creating the "substitution threat": knowing that local potential exists*

Knowing/characterizing/considering extractions in Europe changes the relationship between the customer (European industries) and the supplier (extractive companies often located outside Europe): it is obvious that the supplier envies their commercial relationship with a customer who has a (possible) alternative supply differently, even if the latter is still not well-known in detail. This threat of substitution particularly concerns so-called strategic/critical resources for the European Union. One thinks, of course, of tungsten, which could be (re)exploited in France (Vic 2017).

4.4. Decrease in sites/quantity of operations

The arguments described above have been confronted with several counter-arguments that plead for a reduction, or even a halt, in the short term, of extractions in the Walloon and European subsoil. These arguments can be articulated around two main axes: lack of local skills and, above all, the Not In My BackYard (NIMBY) syndrome (see section 4.4.2), itself based on a dozen causes.

4.4.1. *Lack of local skills (being addressed)*

The *price/commodity price* rose sharply at the beginning of the 21st century (see Chapter 2 of Volume 1). On the other hand, during the decades from 1970 to 2000, commodity prices slowly declined, limiting the investments of mining companies in the field of geological exploration (the first departments to be restricted, in difficult times, are often those of research and development and prospecting) and of states/regions (geological services, study centers and universities). Even more significant, from our point of view, was the associated human disinvestment: for several decades (at least a generation of specialists in the field), the teaching of geological subjects (characterization, prospecting, extraction methods, processing techniques, etc.) was neglected, even stigmatized, in favor of fields considered more "environmental". During the years from 1980 to 2000, students in Earth sciences were led to be interested in areas considered profitable and favorable in the eyes of the citizenry administrations/politics (redevelopment of polluted sites, hydrogeology, risk management, etc.). Thus, the training of human resources in the mining sector has been neglected for several decades in several European countries (this observation is probably less obvious in English-speaking countries). The consequence has been relentless: when metal prices rapidly quadrupled at the dawn of the 21st century, the sector found itself lacking competent personnel at all levels: within the administration to instruct/manage files, within private companies to prospect/characterize/exploit deposits and within universities/schools to teach/vulgarize these subjects. In Wallonia, and also in France, there were few young personnel "competent" in these subjects at the beginning of the 2000s. The geologists active in the field were close to retirement; a generation was missing.

Over the past 15 years or so, this gap has been gradually filled: French universities, in particular, have recently created several posts dedicated to the general theme of subsoil resources (metallogeny, ores, etc.). Specific programs, sometimes associating several European universities, have been created. As a university professor, this brings the author to an observation that goes beyond the mining sector alone: the university (and teaching in general) must maintain a base of so-called "fundamental" skills, including in fields that, at a given time, do not seem "fashionable" but which will be providential later on. Considered as "has-been" in the 1990s, skills associated with the mining sector have been sought after from the end of the 2000s. However, there is nothing to prevent the sector from partly calling on foreign expertise, including for local sites in Europe.

4.4.2. *NIMBY syndrome*

In opposition to the favorable arguments described above, (groups of) citizens have put forward counter-arguments aimed at diminishing/suppressing extractions in Wallonia. These citizens, who sometimes live near active extraction sites, are particularly motivated by the NIMBY syndrome. This syndrome, well-known in many fields combining collective and individual interests, is omnipresent in the extractive sector. The maintenance, or even the creation, of new extraction sites in Europe will be based on social acceptability[3].

Therefore, in order to manage this potentially conflictual situation as well as possible, it is useful to ask what the notions of mine/quarry/extraction/ore, etc., represent for the Walloon citizen nowadays. The perceptions described below result from the synthesis of some 20 years of listening to and observing citizens (residents of extraction sites, opponents of new sites, members of the regional and local administration, politicians, journalists, geologists, operators, etc.): they would undoubtedly merit a rigorous multidisciplinary scientific study. Nevertheless, the following inventory, it seems to the author, provides the basis for the public perception of the field of underground mining.

4.4.2.1. *Polluting, noisy and smelly extraction*

The first perception of a quarry/mine for the citizen is certainly its unpleasant character: polluting, noisy and smelly. Like any industrial activity, quarries/mining extraction generate harmful environmental consequences, to different degrees and on different scales: emissions of CO_2 and other gases, discharges into watercourses, blasting, circulation of mining machinery, etc. On a global scale, the rare media images associated with the mining sector are very often negative: cyanide/mercury pollution for gold mining, breaching of bauxite mine tailings pond dikes, etc. For the citizen, the quarry/mine is above all a harmful object!

4.4.2.2. *The destruction of flora and fauna*

Obviously, a quarry (and, to a lesser extent, a mine) alters the flora and fauna in the immediate vicinity of the extraction site. Consequently, many sites are discovered to be of enormous biological/landscape interest during the impact study carried out following extraction requests. The biological interest of quarries is shown, in many sites, through detailed scientific studies and projects: ecological niches, now inaccessible in nature, are regenerated in the quarries (temporary pools, accessible fronts for birds, etc.). Nevertheless, does nature really need quarries to develop and should we not (re)create these niches elsewhere?

3 This *de facto* term is debatable because it implies that the citizen is led to "accept" the presence of mining extractions; one would prefer the term "citizen perception".

4.4.2.3. *The creation of an additional mode of transportation of goods*

An extractive activity generates locally an additional workload (transport of ore and extraction/processing products, including the transport of personnel and that associated with subcontractors). However, certain alternatives, which are increasingly favored, limit this nuisance: underground transportation of ore, transportation of products by rail or river, and telecommuting for some of the personnel.

4.4.2.4. *Alteration of the landscape during operation*

The quarry site (and to a lesser extent the mine site) alters the landscape (see Chapter 3). In Wallonia (and in some regions of France), this is exacerbated by the legacy of coalmining sites that have left obvious traces in the landscape: slag heaps, disused industrial sites and polluted industrial wastelands. With a surface area of 16,901 km^2 for about 150 sites registered in the sector plan, Wallonia contains about one quarry site per 100 km^2: in this context of proximity (population density is quite high in Wallonia with about 214 inhabitants/km^2), the quarry sector tends to be integrated as well as possible in the landscape (Yans 2017).

4.4.2.5. *Limiting the development of other assignments*

The allocation of surface areas in the extraction zone on the sector plan prevents the development of other economic activities, such as agriculture, forestry and industrial clusters. However, for some years now in Wallonia, quarry sites have had to undergo detailed and credible rehabilitation: the quarry site is then (re)transformed into an agricultural site (filling in with arable land), a tourist site (lake in the former quarry) or even an economic and/or residential site.

4.4.2.6. *The creation of landfills after operation*

As mentioned above, a post-extraction ("post-mining") site rehabilitation plan is today a necessity, although it is well-constrained from an administrative and legislative point of view. It has not always been this way: a good number of old quarries were filled in the last century with rather toxic/harmful waste. The sector has indirectly suffered from this citizen observation. In Wallonia, the cost of rehabilitation of formerly polluted sites is enormous, in particular through the SPAQUE (*Société Publique d'Aide à la Qualité de l'Environnement*), a center of expertise in the Walloon region for the management of polluted soils. To date, such practices have been abolished in Wallonia and increased vigilance must be exercised to avoid any deposit of waste in former extractive sites.

4.4.2.7. *The enrichment of mining companies*

The role of geological resources in our daily lives is essential: the manufacture of all products today requires, directly or indirectly, the use of resources extracted from the subsoil. This fact is largely unknown to the public because of:

– little or no secondary school education (see section 4.4.2.10);

– little media coverage.

Not being aware of the impact of geological resources in their daily lives, citizens perceive the mine/quarry only as a tool to enrich powerful mining companies. This perception is exacerbated by old-fashioned references that are well-anchored in our regions, such as *Germinal* by Émile Zola. The extractive companies are clearly profit-making companies. However, generating extractive activities in Wallonia would promote a fabric of reindustrialization, which has often been desired in recent years, including in politics. This perspective would be facilitated by the Belgian nationality of several extractive companies, some of them world leaders in their field.

4.4.2.8. *Unacceptable social/societal conditions*

Mines/quarries have a bad reputation, in social and safety terms. They are perceived as a "cold" place, the site of human disasters that are often covered by the media. In Wallonia (and France), once again, the legacy of coal mines exacts this citizen perception: the mine is inhospitable, a killer. The disasters that have punctuated local coal mining remain in the collective consciousness (in Wallonia, one thinks of the Marcinelle disaster, which killed 262 miners in 1956). "Firedamp" is an image/expression that is still well-known. The disastrous working conditions of the past remain in all Walloon (and Northern European) memories. Silicosis took away grandfathers: this bitter memory is still very present in families! However, working conditions in European mines/quarries have changed a lot since *Germinal* or since the 20th century. Today, mining sites are very sensitive to safety conditions (it is in their own interest) and limit the dangers. Nowadays, and probably increasingly in the future, modern mines are largely controlled from the subsurface: the mine/quarry has become a place of considerable technological development, favoring working conditions that are not comparable to those of the past or those used in certain parts of the world. Moreover, in the collective unconscious, mining/quarrying is associated with sensitive social notions, such as sometimes complicated immigration and rather unsuccessful restructuring in several sectors (coal mining, iron and steel, etc.).

The current disastrous working conditions in the sector in some countries, partly employing a very young workforce, have been widely publicized in recent years. The explicit notion of *blood ores* has been popularized in several books/reports[4]. These conditions, rightly stigmatized by several NGOs, are sometimes nuanced by certain stakeholders in the field, insisting on the needs of local populations who, without these extractive activities, would be very poor. In any case, unacceptable practices are frequently observed in the sector throughout the world (e.g. in the DRC; Tréfon and de Putter 2017). In addition, the exploitation of mining sites often disrupts traditional societal structures at the local level: for example, communities are shattered when mining sites are developed by international mining companies.

These last points, often neglected in the perception of the mine/quarry by Northern European citizens, are *stumbling blocks*, anchored in the collective conscience, which will have to be better defined/dealt with in the future. The amalgamation with working conditions in the current/future mines/quarries in Europe, although absolutely unfounded, are frequent in citizens' consciousness.

4.4.2.9. *Economic and geopolitical stress*

Titles of articles/reports in recent media coverage associated with the extractive sector can generate stress for the citizen: "shortage alert", "the 26 minerals that we will miss", "the beginning of an end" or "geopolitical tensions around the minerals of new technologies"! Citizen perceptions are then fed by stressful, negative connotations, which can lead to geopolitical tensions, even conflicts. For some minerals, throughout the world, the tension is very real: it could be reduced by local prospections/extractions.

4.4.2.10. *Complex scientific jargon in a little-taught field*

The mine/quarry uses complex jargon at the transition between several disciplines (geology, metallogeny, metallurgy, chemistry, economics, etc.):

4 One can cite, without concern for completeness, C. Boltanski's book *Minerais de sang : les esclaves oubliés du monde moderne*, published in 2012.

fault/tectonics, ore/rock, indium/gallium, coltan/cobalt, grade/tonnage, flotation/eutectic, resource/reserve, strategic/tactical prospecting, etc. In this context, the citizen is quickly disengaged from technical discussions and consequently becomes resistant to any decision. Exacerbating this process, the geologist/engineer perhaps neglects the situation of the non-geologist. This observation is fueled by the absence of notions of geology and geological resources in the official curriculum of secondary schools in the French community of Belgium: today, the 18-year-old Walloon citizen receives very little information at school about the omnipresence, origin and stakes of mineral resources in his or her daily life. It is not easy, under these conditions, to start a debate on solid foundations.

4.5. Some levers for action

The Walloon administration refused the request for exploration submitted by Walzinc in 2017 (see above), notably on the basis of citizen refusal, which crystallized during an information meeting held at the town hall of Plombières on October 4 2017, in the presence of Walzinc officials. Very quickly, during this meeting, the local residents, clearly motivated by the NIMBY syndrome, expressed a categorical refusal for any new extractive industrialization in the region. The mayors of the communes concerned and the regional administration followed this citizen refusal. This experience, among others, has served as a basis for the solutions envisaged below, with a view to a better objective perception of the mine/quarry by the Walloon (and more generally North European) citizen.

4.5.1. *Responsible extraction*

The harmful practices of the past are to be prohibited definitively: the citizen does not tolerate any slips of any nature whatsoever. Moreover, extraction limited to essential sites (in geological, environmental, social and economic terms), with increased consideration/attention to media-sensitive sites, seems to be the only feasible approach today.

4.5.2. *Popularizing*

Popularizing the large contributions of the extractive industry in the daily life of the citizen (i.e. to make people aware of the importance of the subsoil resources in our lives) is a necessary but not sufficient condition:

– it is necessary because if citizens do not perceive the global interest of the extraction of the resources of the subsoil, it will be absolutely impossible to envisage a serene discussion;

– it is not enough because the global approach to the need for underground resources in the daily life of citizens inevitably comes up against more local considerations.

This time-consuming work of popularization can be carried out in several ways: relaying in the media, conferences/extension articles, teaching in secondary schools, etc. The role of federations (FEDIEX – the Belgian Federation of Extractive and Processing Industries for Non-Energetic Rocks – in Belgium, the MIF in France) and learned societies (*Geologica Belgica* in Belgium, *Société de l'industrie minérale, Société géologique de France* in France) is, from this point of view, fundamental. The role of universities is also essential. In recent years, several initiatives in this direction have been carried out at several scales. An ever-increasing effort is required it seems.

4.5.3. *Strengthening the administration and defining a clear public strategy*

In Belgium, the regions manage the subsoil. A few years ago, the public service of Wallonia set up a Geological Survey of Wallonia[5] (*Service géologique de Wallonie*, SGW). However, the human resources of this SGW remained insufficient for several years. Recently, the public service of Wallonia has endorsed the commitment of several geologists as reinforcements within the SGW. A strong administration is needed to address, in a competent and credible manner, the various issues related to the subsoil, including mineral resources. A strong administration will then support policy at all levels (municipal, provincial and regional).

Citizens and decision-makers are calling for a clear public strategy in this area (Yans 2015). In parallel with the strengthening of human resources, the SGW must develop a precise decision-making framework in this area, in order to:

– avoid the processing of administrative files sometimes taking a decade, a timing that is incompatible with the investment periods of extractive companies;

5 It should be noted that there is also a Geological Survey of Belgium, located at the Royal Belgian Institute of Natural Sciences (Brussels).

– prevent every decision from being systematically challenged, rightly or wrongly;

– guarantee general interest.

It is a question of tending (current work is moving in this direction) toward a true *management of the geological resource in Wallonia.*

4.5.4. *Consulting*

Over the past 25 years, we have seen that operators have changed their priorities in managing their quarries/mines. In the past, the priorities revolved around three themes:

– the geological resource itself (nature, volume and geometry of the ore);

– the human resource (staff);

– extraction/treatment costs.

Concerns related to public perception of the extraction site were limited to a few meetings with local elected officials, and even some local residents, and sponsorship of a few local activities (sports club, annual festival, etc.). The role of local residents, and citizens in general, was virtually ignored. *Let's live happily, let's live hidden* seemed to be the motto of the extractive companies. Today, the situation is quite different: the perception of the citizen is a key – if not the main – concern of the operators. The concept of sustainable development is no longer based solely on three pillars (social, economic and environmental) but also on a fourth pillar: citizen participation. Transparency, active communication and openness to the outside world are now the keywords of the sector with regard to citizen perception. Coercive measures from local/regional authorities, which are sometimes necessary, should be considered as a last resort. It is obvious that such a process of permanent consultation is time-consuming and expensive for extractive companies. For several years now, structures (especially at the staff level) specifically dedicated to this task have been set up within the extractive companies and FEDIEX. It seems that this effort should be further intensified in the years to come.

4.5.5. *Collaborating (private–public)*

The paths described above have been facilitated by a real collaboration between private companies and public sectors (administrations, research centers and universities). In the citizen's perception, the extractive company is *de facto* a suspect. In this landscape, the university is perceived as neutral, disinterested, competent, *ultimately* credible and reliable. It is up to academics to better engage in their mission of service to the community (this book is an example of this), in addition to their teaching (especially geological, mining and economic notions associated with the quarry/mining sector) and research missions. It is up to the private companies to initiate/continue collaborations in this sense; their future survival/development depends on it.

4.6. Conclusion

Starting from the (debatable) postulate based on future growing needs for non-renewable geological resources, the subsoil of Wallonia could be further exploited for the supply of certain non-metallic materials, on the one hand, and for the return of metallic extractions on the other hand. The resources exists extraction is technically possible. Local mining would favor a short circuit, limit otherwise deplorable practices and stimulate, to a certain extent, local economic activity. However, local citizen resistance is growing. Will the Walloons, like other Europeans, be capable of (mis)managing future local extractions, which will generate nuisances that would certainly be contained but very real (environmental, noise, visual, social, societal impacts, etc.)? Working conditions in mines and quarries have changed since *Germinal*. The challenge is a major one since it is a question of aligning four *a priori* contradictory objectives: rational use of natural resources, profitable growth for industries, satisfying the increased needs of the consumer and management of the NIMBY syndrome. A clear public strategy, increasingly competent and functional institutions, greater public awareness, "responsible" extractions and private–public collaboration are among the bases of clearly multidisciplinary work, including geology, economics, environment, sociology, geopolitics, history, ethics and philosophy.

4.7. References

Charles, N., Dupuy, J.-J., Christmann, P., Galin, R., Guillon, D. (2017). Industrie minérale et activité minière en France [Online]. Available at: http://www.mineralinfo.fr/sites/default/files/upload/tome_01_industrie_mineraleactivite_miniere_final24032017.pdf.

Choulet, F., Richard, J., Boiron, M.-C., Yans, J., Dekoninck, A. (2019). Distribution of trace elements in the oxidation zone of Zn-Pb supergene deposits: New insights from Belgium MVT deposits. *European Journal of Mineralogy*, 31, 983–997.

Coppola, V., Boni, M., Gilg, H.A., Balassone, G., Léon, D. (2008). The "calamine" nonsulfide Zn–Pb deposits of Belgium: Petrographical, mineralogical and geochemical characterization. *Ore Geology Review*, 33, 187–210.

De Marsily, G. and Tardieu, B. (2018). Stratégie d'utilisation des ressources du sous-sol pour la transition énergétique française. Les métaux rares. Report of the Académie des sciences – Académie des technologies, Paris.

Dejonghe, L. (1998). Zinc-lead deposits of Belgium. *Ore Geology Reviews*, 12, 329–354.

Denayer, J., Pacyna, D., Boulvain, F. (2011). *Le Minerai de fer en Wallonie. Cartographie, histoire et géologie*. Éditions de la région wallonne, Liège.

European Commission (2018). EIP on Raw Materials, Raw Materials Scoreboard 2018. Publications Office of the European Union, Luxembourg.

Goemaere, E., Baele, J.-M., Dejonghe, L., Tourneur, F., Yans, J. (2020). Les Ressources minérales du Massif ardennais. *Géochronique*, 154, 39–42.

Goffin, V., Evrard, M., Pirard, E. (2015). Critical metals in sphalerites from Belgian MVT deposits. In *Proceedings of 13th SGA Biennial Meeting*, 2, Nancy, 741–744, 24–27 August.

Goossens, P.-J. (2014). Zinc potential in Eastern Belgium. *European Geologist*, 37, 7–11.

Laloux, M., Geukens, F., Ghysel, P. *et al.* (2000). Carte géologique d'Henri-Chapelle-Raeren. Service public de Wallonie, DGARNE, Namur.

Mudd, G.M. (2007). An assessment of the sustainability of the mining industry in Australia. *Australian Journal of Multi-disciplinary Engineering*, 5(1), 1–12.

Nassar, N.T., Brainard, J., Gulley, A., Manley, R., Matos, G., Lederer, G., Bird, L.R., Pineault, D., Alonso, E., Gambogi, J. *et al.* (2020). Evaluating the mineral commodity supply risk of the U.S. manufacturing sector. *Sci. Adv. I*, 6, 8647.

Poskin, E. (2010). Activités extractives en Wallonie : une coordination nécessaire. *Etopia*.

Ruthy, I. and Dassargue, A. (2009). Carte hydrogéologique d'Henri-Chapelle-Raeren – Gemmenich-Botzelaar. Service public de Wallonie, DGARNE, Namur.

Tréfon, T. and De Putter, T. (2017). *Ressources naturelles et développement : le paradoxe congolais*. L'Harmattan, Paris.

Vic, G. (2017). Mines européennes. Quelques exemples de mines de métaux [Online]. Available at: http://www.mineralinfo.fr/sites/default/files/upload/tome_10_mines_europeennes_final24032017.pdf.

Yans, J. (2015). L'Exploitation des ressources non renouvelables du sous-sol dans une perspective de développement durable : la vision du géologue transdisciplinaire [Online]. Available at: http://congrestransitiondurable.org/51469/document.

Yans, J. (2017). Gestion durable des ressources minérales en Wallonie (Belgique) : singularités et pistes de réflexion. In *L'Industrie minière et le développement durable : une perspective internationale francophone*, Rouleau, A., Gasquet, D. (eds). CERM, Université du Québec à Chicoutimi and the Institut de la Francophonie pour le développement durable, Quebec.

Yans, J. and Dekoninck, A. (2016). Ressources du sous-sol wallon : historique, état des lieux et perspectives. *Mines et Carrières*, 236, 5–12.

Yans, J., Dusar, M., Swennen, R. *et al.* (2013). Gas shales in Belgium? University of Namur, Namur.

5

Substitution: Promises, Principles and Main Constraints

Florian FIZAINE
IREGE, Savoie Mont Blanc University, Annecy, France

5.1. Introduction

Along with the income effect, the substitution effect is one of the traditional concepts deployed by economists to better understand the behavior of economic players faced with a change in resource scarcity. Economists have therefore mobilized these tools to evaluate the economic impact of the unavailability of certain resources. In fact, these resources, whether used as factors of production or destroyed as consumer goods, open up the field to trade-offs, particularly through substitution toward other alternative resources.

Substitution, therefore, has nothing specific to the economy of natural resources and has been mobilized on many occasions to study the trade-offs between work and leisure, labor input and capital input (and also land input), etc. The substitution of resources is not specific to the economy of natural resources. In reaction to approaches considered as catastrophic (Meadows *et al.* 1972), it was then used as a standard bearer to relativize the exhaustion or unavailability of certain geochemically rare metals, such as copper (Skinner 1976; Goeller and Weinberg 1978), and of natural resources in general (Simon 1981). In this context, substitution toward other more abundant primary or secondary resources (recycling) is one of the key arguments of the cornucopian discourse.

However, the application of traditional economic tools to the field of natural resources is not without theoretical and practical difficulties. It is this generalization that is probably at the origin of a schism between, on the one hand, a "standard economic science" replicated to the environment and natural resources and, on the other hand, an "ecological economy" that refutes the application of these hypotheses on mainly physical and thermodynamic principles. This split thus opposes an economy of weak sustainability – assuming a strong substitutability between natural capital and technical/human capital – to an economy of strong sustainability (or assuming a weak substitutability of natural capital).

Without excluding the problem of resource depletion, we will not enter into the terms of this debate; first of all, because this chapter is dedicated above all to the practical possibilities and limits inherent in the substitution of mineral resources, and also because the possibilities of specific shortages also exist in the short term and are not part of these long-term considerations. The energy transition will call for more metals in the medium term, but it will not be the only one to do so: this is also the case, for example, with the digital transition. There could therefore be conflicts of use over some common uses of resources, as was the case between agriculture and transport for first-generation biofuels. Determining the extent to which each sector has substitution possibilities and can therefore give up one resource for another then appears to be essential.

Substitution has been the subject of numerous applied academic studies over the last half-century, so we propose here to undertake a synthesis applied as far as possible to mineral resources for energy. We first lay out the main economic bases of substitution and then we analyze the different forms it can take. We then identify the major constraints highlighted in the literature. Finally, we conclude with other aspects more specific to mineral resources that may impact the effectiveness of substitution as a lever of action.

5.2. Main economic foundations of substitution

First of all, it should be remembered that economic agents do not consume and therefore do not demand natural resources *per se*. In other words, we do not consume copper or even copper cables, but rather a material substrate that can provide us with a high level of electrical and thermal conductivity to provide communication and energy transmission services. Thus, it is a derived demand, which economists model from the notion of the demand curve. We will now look at the concept of the demand curve and the main mechanisms interacting with it.

5.2.1. *The demand curve*

The demand curve is a conceptual tool describing the relationship between the quantity demanded of a resource and its price over a particular period of time, all other things being equal. In general, it is assumed in standard microeconomics that the demand curve is a decreasing function of price, that is, that the quantities demanded for a resource decrease as its price increases (although there are special cases where this is not the case). This decrease in the quantity demanded following a price increase can be explained by two phenomena: a *substitution* effect, because economic agents then demand less of this resource and more of other resources, and an *income* effect, because this price increase leads to a decrease in the real income of economic agents, who can then consume less of all resources. Given the low importance of the cost of raw materials in the cost of final products, the income effect is often considered negligible and the reduction in demand for a natural resource is then almost entirely attributable to the substitution effect (Tilton 1984).

This response of demand to price can vary according to the resources and above all according to the technical possibilities of substitution. Economists use the notion of price elasticity to measure this response of demand to a change in its price. It is generally accepted that a price elasticity is negative and low (close to 0) if the possibilities of substitution are limited. In contrast, easy possibilities of substitution (immediate and without cost) should result in a high price elasticity, i.e. a strong downward reaction of demand to a change in price. Of course, this price elasticity of demand for resources also varies according to the characteristics of the applicant and the market in which it is present. A study such as that of Nassar (2015) proposes to measure the price response of platinum demand for different sectors (demanders). To do so, he regresses the annual demand quantities of platinum for each sector according to the price of platinum and other controlling factors (GDP, prices of substitutes). In these configurations, he arrives at a price elasticity of demand for platinum of -0.20 for electrical appliances, -0.32 for chemicals, -0.45 for the oil sector (refining), -0.70 for jewelry and -2.68 for the financial investment sector (platinum is then a financial asset). This means that a 1% increase in the price of platinum should translate (on average) into a decrease in demand of -2.68% in the financial investment sector, but only -0.2% in the electrical applications sector. In the event of a very sharp rise in the price of platinum, one would therefore expect financial investors to throw in the towel much more quickly than the electrical applications industry in the bid for the precious metal.

Another relatively conventional question is that of the identification of substitutes close to the resource as well as the reaction of demand to a change in the price of these alternative resources (for the same price of the resource). We then speak of *cross-price elasticity*. A positive cross-price elasticity implies the presence of a substitute for the resource, whereas a negative cross-price elasticity rather indicates the presence of a complement to the resource. Let us illustrate this in the case of platinum group metals (PGMs). Nassar (2015) also measured this reaction of platinum demand (depending on the sector) to changes in the prices of the closest substitutes. In the case of the jewelry sector, alternatives to the use of platinum can be similar to Au-Cu-Ni-Zn or Au-Pd-Ag alloys, with nickel and palladium providing the bleaching action in the absence of platinum. However, the study shows that, although statistically significant, the response is low (0.21 for nickel and palladium). In other words, when the price of alternative materials to platinum drops by 1%, it only marginally affects the demand for platinum (0.21% drop). One can therefore suspect that the jewelry sector does not consider these alternatives to be sufficiently effective. On the other hand, gold in the financial investment sector has a much higher cross-price elasticity with the demand for platinum of around 2.93. These highly applied studies are thus very valuable for a better understanding of the behavior of the sectors that use a raw material and the substitution possibilities offered to each sector.

5.2.2. *The horizons of substitution: short, medium and long term*

However, the price response of the demand for a resource is far from being a temporal constant, and it is also considered that it can vary according to the time horizon studied. As Tilton (1984) rightly points out, occasional shortages imposed by embargoes, cyclical increases in demand, strikes and natural disasters often result in considerable price increases without, however, causing major immediate substitutions. This can be mainly explained by the time horizon studied. It should be remembered that in order to substitute one resource for another (copper for aluminum, for example), producers, if they wish to react in the very short term, need a substitute:

immediately available, at a competitive price, requiring no change in equipment and no new technology. These possibilities exist but are the exception rather than the general case. In contrast, if we look at the medium term, the price response may be greater because producers can change their equipment to use another resource. Finally, in the very long term, the lack of substitutes can sometimes be compensated for by investment in research and development and potential new technologies. However, given the time required for these developments, these substitution possibilities are rarely deployed to deal with short-term shortages.

5.2.3. *The shortcomings of the classical demand curve*

The demand curve concept approach, however, is far from being a panacea because of its strong underlying assumptions: reversibility, continuity and intertemporal stability. First of all, this concept implies a notion of reversibility: although it is claimed that following an increase in the price of a resource, its demand falls, this does not exclude a return to the initial equilibrium if the price returns to its previous level. In the medium to long term, and especially in the event of a significant price increase, this hypothesis is unlikely to be realized because of the potential definitive destruction of (part of) the demand, as producers have switched to more efficient equipment or technologies. In this case, we are no longer witnessing a shift along the demand curve but a shift in the demand curve.

To continue, the hypothesis of continuity is also debatable and one can very well imagine "jumps" in demand in the face of significant increases or decreases in the price level: extrapolation from average points then has its limits. Finally, the intertemporal stability of demand is also unconvincing because of the factors mentioned above, and also because it is in any case very difficult to be sure of this "all things being equal" when passing into the field of measurement.

5.3. Elements, components, systems: what are we really substituting?

Far from being limited to a substitution from element to element, substitution is a much more general phenomenon that needs to be better understood in order to assess the potential of this lever of action. To better explain these scales in substitution, authors such as Rankin (2011), Fizaine

(2014) and Smith and Eggert (2016) have used Altenpohl's (1980) hierarchy or a categorization very close in spirit to the five types of substitution of Tilton (1984), which we will present. These scales do not explain everything, and it is also appropriate to consider the types of resources to which we are making a transition. Here again, some authors have proposed normative paths for the future in order to ensure greater sustainability of our societal systems.

5.3.1. *Altenpohl hierarchy and principal forms of technical substitution*

The Altenpohl hierarchy aims to define the different scales of substitution. Thus, the author distinguishes seven different levels of substitution:

1) a change in purpose or personal/social value resulting in a modification of the application;

2) a change in strategy to achieve objectives;

3) a change of the system via technical means allowing the implementation of a strategy;

4) a change of subsystem through a modification of the system design;

5) a component change through a design change;

6) a material change for a specific component;

7) a change in the way the resource is produced.

As can be seen, all levels affect the consumption of raw material and in particular the specific type and amount of metal called up. The first types of substitution are not only very impactful but also very difficult to put in place (very long-term modifications). Although not in direct correspondence, Tilton's (1984) five surrogate factors are very close in spirit:

– element-to-element substitution;

– substitution of other factors (labor, capital and energy) for the resource;

– substitution of quality type for the resource;

– inter-product substitution;

– technological substitution.

As can be seen, the first type of Tilton substitution corresponds to level 6 of the Altenpohl hierarchy, while it is difficult to hang up the other types of substitution at specific levels. For example, technological substitution can just as easily trigger a change of component, subsystem, system or strategy. Concrete applications of these concepts have been made in favor of tin (Tilton 1984), metals for greenhouse gas reduction in transportation (Fizaine 2014) and neodymium-based permanent magnets in wind power (Smith and Eggert 2016).

While Tilton estimates that element-to-element and technological substitutions are the most frequent in the case of tin, Smith and Eggert (2016) show that in the face of an increase in the price of rare earths, technological (modification of the arrangement of the elements without altering the quality), quality (grade-for-grade substitution) as well as system and subsystem (type of energy or wind power with gearbox) substitutions were mainly used. In any case, we can see that the substitution effect varies according to place and time depending on the actors, the interactions with the market structure and the availability of alternative technologies and substitutes at the different scales of substitution. So the trick is in the details. However, although particularly relevant from a technical point of view, these scales do not address the question of the long-term sustainability of our societies.

5.3.2. *Normative substitution: what to substitute for?*

This is why some authors have gone further by proposing sustainability scenarios based essentially on the lever of substitution. Faced with the economic depletion of geochemically rare mineral resources such as copper, tin, gold, platinoids or other minor metals, academics have evoked three major families of substitution.

For example, Skinner (1976) and also Goeller and Weinberg (1978) estimate that we will eventually tip over into a new Iron Age. They thus evoke a massive future substitution of geochemically rare metals (not very present in the Earth's crust) by abundant metals, that is to say essentially iron and aluminum (but one also evokes titanium and magnesium). One thus speaks here of intra-mineral substitution to compensate for the economic unavailability of metals present in very small quantities in rare and concentrated deposits.

Other authors consider even broader substitutions, such as Daly (1991), who points out that the rate of use of renewable resources should not exceed their regeneration rate and that the rate of use of non-renewable resources should not exceed the rate at which renewable substitutes for these resources are developed. The author therefore here proposes a shift from an economy based on non-renewable stock resources to one based on renewable flow resources.

Finally, the last category of substitution goes back to Hartwick (1977). He recommends investing the rent from the extraction of declining natural resource stocks in technical and human capital. The increase in technical and human capital would then take over to compensate for the possible decline in consumption that could result from the economic or physical exhaustion of the former.

If these three ranges of substitution are intended to compensate for the possible future scarcity of mineral resources, it is also easy to see the pitfalls of these substitution solutions as the only lever of action. First, no economic system can be sustainable in the very long term (some speak of 1,000 years) if the extraction of finite resources (renewable or not) increases. Exponential consumptions reduce almost any gigantic stock value to some hundreds of years of consumption (at best). The substitution of one element by another saves time but does not solve the problem of the increasing pressure that the anthroposphere puts on the biosphere. We are simply shifting pressure from one point to another (Kleijn *et al.* 2011). Second, far from moving to an economy based on renewable flows, our economies are doing exactly the opposite (and it is getting worse), i.e. they are increasingly dependent on stock resources (Krausmann *et al.* 2009).

Moreover, most renewable resources are already very well exploited or even overexploited (Allwood *et al.* 2011), so it is difficult to see how this mino-total share of resource withdrawals could replace the majority non-renewable share. Finally, while it is theoretically possible to conceive of the total substitution of a metal such as copper by technical progress (and some people still doubt it), it is quite different if it is a question of substituting all metals, mineral resources or even natural resources by technical and human capital, which moreover also require natural resources for their growth. If this hypothesis were justified, it would mean that the quantity of cakes could be increased indefinitely thanks to the improvement of ovens, machines and bakers' diplomas while constantly reducing the quantities of flour, water and energy.

We can see here that substitution has its limits and that these limits are not limited to those mentioned above. We will now look at the main obstacles identified in the literature.

5.4. The main obstacles to substitution

Indeed, well-documented studies are now available on the main obstacles to substitution, including Messner (2002) on copper and aluminum, Nassar (2015) on platinoids and the more general article on resources by Ayres (2007). These barriers can be grouped into a few broad categories: technical, economic, those related to physical availability, historical, cultural and regulatory.

5.4.1. *Technical obstacles*

To begin with, substitution often comes up against technical constraints; that is, in the best of cases, the potential substitutes of a resource are rarely perfect substitutes in the sense that they would obtain similar performances for all the characteristics fulfilled by the resource subject to substitution (optical, mechanical and thermal performances, conductivity, various resistances, etc.). In this case, one is exposed to a decrease in performance following substitution. Sometimes this is conceivable and directly manageable (e.g. catalysts not using precious metals), other times it is formally excluded because of the central role of the performance and quality of the product or service incorporating the element (alloy in aerospace or nuclear industry). According to Messner (2002), it is this potential aversion to performance loss that could sometimes prevent producers from switching to a viable alternative substitute.

In still other cases, there is simply no substitute identified as having the desired characteristic (Ayres 2007; Nassar 2015), for example, automotive catalytic converters (based on rhodium, palladium and platinum) or the use of chromium in industrial machinery. The metal can be technically substitutable in certain uses (with or without technical loss) and on the contrary have no substitute in certain other functions. The study by Graedel *et al.* (2015) explored the range of substitutes for 60 elements in more than 80% of their mass market outlets. This titanic work led them to observe that certain metals are globally substitutable (aluminum and zinc), while others are much less so (copper, chromium, manganese and lead) or even potentially not at all (rhenium, PGM and REE). Then, they also noted that the substitution of these special metals is possible through metals from the same group, as found also by Fizaine (2013). Moreover, while it is true that there are possibilities of substitution between natural resources for most of the elements used in the construction and clothing sector, we do not know of any conceivable substitute for the whole category (Ayres 2007). The materials used today are the result of an optimal selection that has taken place over decades or even hundreds of years, so it is rather unlikely that we can substitute traditional materials with new elements. Cotton, paper and glass (for windows) are perfect illustrations of this optimal selection.

However, even the presence of an adequate technical substitute does not necessarily make the substitution certain, as we will see with the economic constraints.

5.4.2. *Economic obstacles*

There are also many economic obstacles. As Nassar (2015) reminds us, the relative price between two substitutable resources may not be favorable to trigger substitution despite technical feasibility. Messner (2002) furthermore believes that this favorable relative price must be stable and fluctuate little in order to convince the producer to initiate the substitution process. More generally, the author considers that even a favorable relative price ratio may not lead to a change in the allocation of resources if there are prohibitive switching costs. We return here to the short/medium/long term distinction made by Tilton (1984). In reality, most substitutions require readjustments in equipment and technology, but Messner shows that other benefits are also lost as a result of substitution:

– the benefits obtained in terms of knowledge on and the learning effect of the old element;

– the existence of cooperation networks with suppliers and customers woven through the former element. Reclaiming these advantages therefore requires time and money, while it would also be necessary to assume;

– the potential costs of capital reorganization (Nassar speaks of “sunk” investment costs).

Consequently, since the price of the material is not very significant given its weight in the cost of the final product compared to other factors of production (capital, energy, labor, etc.), the producer often chooses to minimize the additional costs associated with material substitution for these three other factors of production. In most cases, this leads the producer not to consider intra-material substitution. For Messner (2002), these different factors have favored the maintenance of copper in different uses despite a higher relative price than aluminum. Beyond these costs of change, there is also a *path dependency*, which means that current technical choices are constrained by their past history, which would have played out in favor of copper.

In other situations, the use of more expensive metals can significantly reduce other variable costs (e.g. through the ability to operate at lower pressures and temperatures). As a result, the relative price between two resources may be favorable for their substitution without the full cost–benefit analysis of the operation.

Second, there are subsidies inherent in certain resources, products or technologies that do not cover possible substitutes. This leads to market distortions that also distort incentives, as in the case of dental work with palladium-gold components in Japan (Nassar 2015). Scrapping premiums also have a strong influence on the prices of recycled materials and their primary substitutes.

To conclude, the price effect would have little influence on intra-material substitution and would suffer from a very strong inertia that could last beyond the decade (Messner 2002).

5.4.3. *Barriers related to the physical availability of the resource*

Without being directly economic constraints, the risks related to the physical availability of a resource can also be a barrier to a certain number of substitutions. For example, a resource that is likely to suffer from a high risk of supply disruption (wars, embargoes, etc.) may be excluded from technological choices despite unbeatable technical performance and a very favorable economic cost. The case of electric motors using or not using rare earth metals is emblematic. While many hybrid cars included rare earth metals in their batteries (NiMh) or motors (permanent magnets) at the end of the 2000s, the rare earth metal crisis and its negative impact in terms of image very quickly convinced manufacturers to do without this type of metal. Today, the Renault Zoe and Tesla cars do not incorporate any rare earth metals, despite their technical superiority.

In still other situations, it is the physical availability of the resource that is insufficient in view of its needs. For example, according to Nassar (2015), although iridium can technically substitute for platinum in the glass manufacturing sector, it is highly unlikely that this substitution will occur given the large quantities of platinum used. Indeed, despite a favorable relative price in 2010 ($35,000/kg vs. $48,000), iridium was produced in much smaller quantities (10 tons vs. 183 tons per year). Overall, the author believes that this situation prevents the use of minor metals that could substitute for certain other major metals (Fe, Cu, Al, etc.).

Additional complexity, co-production or by-production can greatly reduce the interest in substitution (Graedel 2011; Fizaine 2013). As noted above, alternative substitutes for a metal are frequently other metals with similar characteristics. Unsurprisingly, they are often found associated in the same deposits (e.g. PGMs and rare earths). Switching from one metal to another exposes the consumer to an increase in the price of the substitute without a strong reaction from the joint supply of metals (low price elasticity of co-products and by-products). The unsuccessful switch from platinum to palladium has thus been well documented. Other rationing situations are sometimes even more difficult to perceive. Verhoef *et al.* (2004) have demonstrated that the substitution of lead in solders has rapidly led to the rationing of these potential substitutes (bismuth and silver). This is mainly explained by the frequent primary product status of lead for these two metals as a by-product. Reducing the demand for lead reduces the supply of lead and thus the supply of these substitutes. Similarly, the substitution of major primary (mined) metals by recycling may reduce the availability of primary minor metals that are predominantly by-products of the former (Hagelüken and Meskers 2010). This can again cause problems if these primary minor metals are frequently associated with recycled major metals and the situation becomes more widespread.

5.4.4. *Cultural and historical barriers*

The last potential obstacles to substitution, cultural and historical resistance can also sometimes strongly reduce the opportunities for substitution. Indeed, as Nassar (2015) reports, these factors exist in particular in the jewelry sector, where the selection of metals was delimited a long time ago and where certain elements are an integral part of the culture, as is the case in India with gold. Similarly, platinum also appears to be difficult to substitute in this same sector, which is reflected in a very low price elasticity of demand. Other sectors also seem to be affected by these resistances based on historical use, as is the case with platinum-based non-stick coating technologies.

This impact of culture not only constrains substitution, but it can also sometimes encourage it. For example, Carmona *et al.* (2017) show that plastic in the 1950s had a better image than ceramic dishes (quieter, more hygienic, softer and lighter). Today, by performing the same functions, it is perceived as tasteless, hazardous to health and unhygienic. The relationship between a resource and the culture is therefore malleable over time and partly determines the markets.

5.4.5. *Regulatory barriers*

Finally, an important source of forced substitution or, on the contrary, the prevention of substitution lies in the constraints imposed by regulations and whether they are motivated by the environment, safety or health.

In particular, the situation of UK occupational safety standards that require demolition workers to work mainly on the ground has reduced the availability of steel for reuse and therefore the substitution of other materials by steel (Allwood *et al.* 2011). Some standards also prescribe mixtures or compositions rather than prohibit a level of performance, which may inhibit reuse/substitution to prescribed materials (see Allwood *et al.* (2011) for the case of concrete barriers in the United States). Similarly, this type of effect is largely responsible for the substitution of several heavy metals (mercury and lead) in products and possible second-round effects on other accompanying metals[1] (Nassar 2015). At other times, these same standards do not prescribe a material's use but strongly restrict it. In particular, Tilton (1984) explains that the restriction on the use of tin cans in the U.S. domestic market during military emergencies quickly led to high material efficiency and a reduction of tin in alloys. Allwood *et al.* (2011) describe a general lack of certification of recycled materials because standards do not consider reuse, again preventing substitution between primary and secondary resources.

1 As new components often require a complete reorganization after removal of one of the heavy metals, other previously associated metals are substituted in turn.

5.5. Other aspects to be taken into account

These large groups of constraints do not cover all the factors at play in the substitution process, first because it can also result from the strategic behavior of economic agents concerned about preserving their outlets, and second because the economic substitution of mineral resources raises the question of their ecological substitutability.

5.5.1. *Impact of competition and industrial strategies*

As mentioned above, the hypothesis of demand reversibility appears to be very theoretical, especially for very high increases and levels of raw material prices. In other words, it is possible that in the event of a sharp rise in the price of a metal, part of the demand for it may definitively disappear (due to substitution or material efficiency at the level of demand) even if the initial price is restored. This thus opens up the possibility for producers to develop price control strategies, particularly during a boom period (Tilton 1984). Indeed, to prevent a definitive loss of demand, producers may be reluctant to increase their price when the market equilibrium requires it. Some analyses of the oil market put forward this type of argument by evoking the existence of a price ceiling imposed by the threat of a potential substitute (*coal to liquid* and *gas to liquid in* particular). One can therefore imagine that metal producers have the possibility of playing on their margins to transmit or not transmit price increases. The study by Bakhat and Würzburg (2013) corroborates this type of adjustment with a threshold effect. Thus, aluminum, copper and nickel prices are impacted by the price of oil via production costs and transport costs. However, metal producers do not react in the same way according to the magnitude and sign of the oil price variation, thus underlining the impact of strategies on behavior.

On the other hand, in their analysis of the strategies of permanent magnet producers in the face of the rare earth metal crisis, Smith and Eggert (2016) have shown that the absorption or pass-through of increases in the price of a resource is also part of the portfolio of strategies (in addition to substitution). More specifically, their panel of experts and industrialists believe that although the absorption of price increases (by reducing margins) may have existed during the crisis, this was not the dominant response and most of the increase was passed on to intermediate users (those who buy permanent magnets).

Of course, these case studies do not pretend to be generalized, but simply illustrate the possibility or not of a lack of transmission of price increases depending on the substitution choices of downstream users and the possibilities of reducing the margins available to producers.

5.5.2. *Is economic substitution also an ecological substitution?*

One of the issues directly related to the substitution lever is that of ecological substitution. In other words, does economic substitution (that regulated by markets) necessarily translate into ecological substitution, which would be less costly from an environmental point of view?

Here again, there are major controversies and the debates are often decided on the basis of particularly detailed lifecycle analyses. Indeed, Gutowski *et al.* (2013) observed that the substitution of low energy-intensive materials by high energy-intensive materials has been favored. In parallel, Carmona *et al.* (2017) document this effect for the automotive industry, where the amount of unalloyed steel and iron was reduced by a third, while the amounts of high-performance steel, magnesium, aluminum, plastics and composites were multiplied by 5, 10, 4 and 2, respectively. All of these materials are more energy-intensive than their precursors[2]. The paradox becomes even more pronounced when we also notice that the correlation between the amount of energy incorporated in the materials and their price is very strong (R = 0.96). This means that these major changes are not only more costly from an energy point of view but also less efficient from an economic point of view (more expensive). So why are these substitutions made? Could not we save money by replacing very energy-intensive materials with less energy-intensive materials? In reality, there are strong technical constraints on the products and a low energy weight of the materials in the product lifecycle[3]. It therefore seems more interesting to have lighter, more complex and expensive materials (both in monetary and energy terms) at the construction stage, but which guarantee reduced energy consumption during the use phase (especially for buildings and vehicles). These substitutions are therefore a win-win situation from an economic and energy point of view and probably also if we look at the impact in terms of greenhouse gases

2 At the same time, the average weight of vehicles in the UK went from 995 to 1,321 kg between 1975 and 2012.

3 Most of the authors mention ranges between 5% and 20% of the total energy consumption of products (manufacturing, use end of life).

(GHGs). In this specific case, the substitution is therefore synonymous with lower impacts in terms of GHGs.

However, it is simplistic to equate the full environmental cost of a material with its energy cost. Limiting the analysis to this aspect means leaving out many factors of the ecological load of products and materials and therefore potential external effects (Smil 2013). Contrary to cost–benefit economic analyses, where all aspects are converted to a monetary scale, lifecycle analyses (LCAs) study the impact according to different criteria (energy, global warming potential, soil erosion, water, acidification and eutrophication potential). A substitution can therefore be a winner on some points but a loser on others. For example, Allwood *et al.* (2011) and also Carmona *et al.* (2017) recognize that although new materials that are more energy-intensive in construction can be a favorable option over the entire lifecycle of products, they are also often more complex and less recyclable. Thus, there is a trade-off between more energy (and GHGs) and less waste, or the reverse. The economist's conventional response would be to recommend monetizing the external effects associated with these two dimensions (GHGs and waste). The other aspects would already be integrated by the markets and thus managed in an optimal way (Soderholm and Tilton 2012).

Consequently, apart from the external effects involving market failures, economic substitutions would therefore be adequate. However, the monetarization of all the external effects associated with the production of natural resources is tantamount to opening a Pandora's box: what about the measurement of these effects? What about the temporal and geographical variability of these measures? The associated uncertainty? To our knowledge, there are no studies that investigate monetizing all (or even a majority) of the external effects associated with the production of metals. However, we cannot make progress without more information on the degree of complete ecological sustainability of the substitutions in progress and the direction that can be given to them via public policies. This is a gap that future studies will have to fill and which will probably show the extent of our ignorance on the subject.

5.6. References

Allwood, J.M., Ashby, M.F., Gutowski, T.G., Worrell, E. (2011). Material efficiency: A white paper. *Resources, Conservation and Recycling*, 55, 362–381.

Altenpohl, D. G., (1980). *Materials in World Perspective*, Springer-Verlag, Berlin.

Ayres, R.U. (2007). On the practical limits to substitution. *Ecological Economics*, 61, 115–126.

Bakhat, M. and Würzburg, K. (2013). Co-integration of oil and commodity prices: A comprehensive approach [Online]. Available at: https://ideas.repec.org/p/efe/wpaper/fa05-2013.html [Accessed on 25 May 2021].

Carmona, L.G., Whiting, K., Carrasco, A., Sousa, T., Domingos, T. (2017). Material services with both eyes wide open. *Sustainability*, 9, 1508.

Daly, H.E. (1991). *Steady-State Economics*. Island Press, Washington.

Fizaine, F. (2013). Byproduct production of minor metals: Threat or opportunity for the development of clean technologies? The PV sector as an illustration. *Resources Policy*, 38(3), 373–383.

Fizaine, F. (2014). Analyses de la disponibilité économique des métaux rares dans le cadre de la transition énergétique. PhD Thesis, University of Burgundy, Dijon.

Goeller, H.E. and Weinberg, A.M. (1978). The age of substitutability. *The American Economic Review*, 68(6), 1–11.

Graedel, T.E. (2011). On the future availability of the energy metals. *Annu. Rev. Mater. Res.*, 41, 323–335.

Graedel, T.E., Harper, E.M., Nassar, N.T., Reck, B.K. (2015). On the materials basis of modern society. *PNAS*, 112, 6295–6300.

Gutowski, T.G., Sahni, S., Allwood, J.M., Ashby, M.F., Worrell, E. (2013). The energy required to produce materials: Constraints on energy-intensity improvements, parameters of demand. *Phil. Trans. R Soc. A*, 371.

Hagelüken, C. and Meskers, C.E.M. (2010). Complex life cycles of precious and special metals. In *Linkages of Sustainability*, Graedel, T.E., van der Voet, E. (eds). The MIT Press, Cambridge.

Hartwick, J.M. (1977). Intergenerational equity and the investment of rents from exhaustible resources. *American Economic Review*, 67, 972–974.

Kleijn, R., van der Voet, E., Kramer, G.J., van Oers, L., van der Giesen, C. (2011). Metal requirements of low-carbon power generation. *Energy*, 36(9), 5640–5648.

Krausmann, F., Gingrich, S., Eisenmenger, N., Erb, K.-H., Haberl, H., Fischer-Kowalski, M. (2009). Growth in global material use, GDP and population during the 20th century. *Ecological Economics*, 68, 2696–2705.

Meadows, D.H., Meadows, D.L., Randers, J., Behrens III, W.W. (1972). *The Limits to Growth*. Universe Books, New York.

Messner, F. (2002). Material substitution and path dependence: Empirical evidence on the substitution of copper for aluminum. *Ecological Economics*, 42(1–2), 259–271.

Nassar, N.T. (2015). Limitations to elemental substitution as exemplified by the platinum-group metals. *Green Chemistry*, 17, 2226–2234.

Rankin, W.J. (2011). *Minerals, Metals and Sustainability: Meeting Future Material Needs*. CRC Press, Boca Raton.

Simon, J.L. (1981). *The Ultimate Resource*. Princeton University Press, Princeton.

Skinner, B.J. (1976). A second iron age ahead? *American Scientist*, 64(3), 258–269.

Smil, V. (2013). *Making the Modern World, Materials and Dematerialization*. John Wiley & Sons, Hoboken.

Smith, B.J. and Eggert, R.G. (2016). Multifaceted material substitution: The case of NdFeB magnets, 2010–2015. *The Journal of The Minerals, Metals & Materials Society*, 68(7), 1964–1971.

Soderholm, P. and Tilton, J.E. (2012). Material efficiency: An economic perspective. *Resources, Conservation and Recycling*, 61, 75–82.

Tilton, J.E. (1984). Material substitution: Lessons from the tin using industries. *Resources for the Future*, RR-84-9.

Verhoef, E.V., Dijkema, G.P.J., Reuter, M.A. (2004). Process knowledge, system dynamics, and metal ecology. *Journal of Industrial Ecology*, 8, 23–43.

6

Resource Consumption and Decoupling

Thierry LEFÈVRE
Center for Research on Advanced Materials (CERMA), Laval University, Quebec, Canada

6.1. Introduction

The extraction of natural resources, their combustion or consumption, their processing to produce goods, as well as the use, maintenance and disposal of these goods at the end of their useful lifespan are behind environmental degradations that have local effects and, more generally, modify the processes that regulate the planet (Mackenzie 2011; UNEP 2019). In addition to the deterioration of the ecosphere, social consequences (Sachs 2015) also threaten the quality of life, human health and the proper functioning of societies. The excessive use of non-renewable resources is also detrimental to their intergenerational availability.

In response to the environmental threat, scientists have established limit values for critical parameters characterizing terrestrial processes, which they have called "planetary boundaries" (Rockström *et al.* 2009; Steffen *et al.* 2015). These parameters represent climate, biodiversity and ecosystem services, the nitrogen and phosphorus cycles, ocean pH, the integrity of natural territories, freshwater use, stratospheric ozone and pollution. Planetary boundaries represent ecospheric values that should not be exceeded to ensure safe land use, comparable to that experienced by *Homo sapiens* during the Holocene.

Mineral Resources Economics 2,
coordinated by Florian FIZAINE and Xavier GALIÈGUE.

Planetary boundaries are intended to preserve the Earth's state by setting limits on the *consequences of* human activities. Focusing on the *causes* of environmental degradation, scientists set limits on the global environmental footprint[1] in 2014 (Hoekstra and Wiedmann 2014). This footprint is divided into four categories: the ecological footprint (area needed for food, housing, production/consumption and discharges), the carbon footprint (CO_2 emissions), the water footprint (freshwater use) and the material footprint (use of natural resources). These footprint limits aim to quantify the maximum use of resources and territories.

These requirements are all the more pressing as resource exploitation is increasing at an accelerating rate (UNEP 2016; Krausmann *et al.* 2018). Global resource extractions have increased from ~7 gigatons per year (Gt/year) in 1900 to ~82 Gt/year in 2015 (Krausmann *et al.* 2018). A particularly marked acceleration occurred at the turn of the 2000s, such that the growth rate of global use increased from 1.8% between 1973 and 2002 to 3.3% between 2002 and 2015 (UNEP 2016; Krausmann *et al.* 2018). The growth in global extractions between 2002 and 2015 was even faster than the growth in gross domestic product (GDP) or world population (UNEP 2016). On average, each person used 12.1 t of resources per year (12.1 t/year.cap) in 2015 compared to 9.3 t/year.cap in 2002 (Krausmann *et al.* 2018) and 4.6 t/year.cap in 1900 (Krausmann *et al.* 2009, 2018). The total waste stream increased from 14 Gt/year in 1900 to 111 Gt/year in 2015. Our civilization has thus released a total of 2,500 Gt of waste into the environment in 115 years (Krausmann *et al.* 2018).

This acceleration in extraction particularly concerns non-renewable resources: the use of minerals has increased by a factor of 34 over the 20th century, notably due to urbanization (Krausmann *et al.* 2009, 2018). At the same time, the use of metals has increased by a factor of 27 (Krausmann *et al.* 2009). Between 2000 and 2010, the rate of growth in extractions accelerated except for renewable resources. The average extraction rate for this period was 5.3% for non-metallic minerals, 3.5% for metals, 2.9% for fossil fuels and 2% for biomass (UNEP 2016).

The exploitation of natural resources contributes greatly to the functioning and development of nations. Material consumption increases especially in phases of development such as those experienced by OECD

1 The concept of a "footprint" will be further elaborated below in the context of resource use assessment.

countries after World War II and China in recent decades. As the use of resources increases with GDP, and therefore with wealth and economic growth, so-called "mature" societies consume and reject more materials than poor countries. Finally, the fraction constituted by mineral resources increases in relation to biomass during industrialization.

Excessive material use may not only lead prematurely to the depletion of non-renewable resources, but also to a shortage of renewable resources if the rate of extraction exceeds their replenishment capacity. Resource withdrawals also have direct environmental impacts, such as land degradation and pollution of soils and aquifers. Finally, resource flows have innumerable indirect impacts because they support human activities and their environmental consequences (combustion of fossil fuels, urbanization, industrial production and consumption, etc.).

Environmental degradation has already reached such a level (Steffen *et al.* 2015; Ripple *et al.* 2017; UNEP 2019) that many consider the consumption of resources to be excessive, but putting a figure on "reasonable" consumption is not trivial. This exercise has been attempted and suggests that global resource use should be limited to 8 t/cap (Dittrich *et al.* 2012). A future convergence of emerging countries with the current material footprint of industrialized countries accentuates the threat to the ecological balance and reinforces the idea that the human environmental footprint should decrease. In other words, it is imperative to dematerialize the economy. Another way of expressing this need is as decoupling economic growth from resource use. This idea was part of the OECD's concerns in the early 2000s and was strongly supported by the UNEP in 2011 (UNEP 2011).

Decoupling is a goal that often underpins sustainable development and the green economy. This vision is based on the assumption that economic growth can continue while reducing the environmental footprint. Thus, the quantities of resources used are compared to GDP, even though this criterion does not necessarily guarantee human well-being. Decoupling also requires finding indicators to adequately measure the use of natural resources in a context where the economic system is globalized and marked by innumerable trade exchanges. This issue is all the more important as use differs greatly between rich and poor countries, sometimes by a factor of 10 (Steinberg *et al.* 2010; Dittrich *et al.* 2012; UNEP 2016). These arguments lead to the issues of equitable distribution and use of natural resources and the success, or otherwise, of decoupling by different countries.

This chapter is devoted to decoupling and the notions surrounding this concept. It begins with a description of material flows (inputs and outputs) and their accumulation in the form of manufactured capital within a society, a set of indicators characterizing its socio-economic functioning. This is followed by an overview of the different ways of quantifying national resource use from the point of view of exports and imports. Decoupling is then defined and an assessment is made of whether this objective has been achieved and, if so, by whom and under what circumstances; if not, it is a question of what the limiting factors are. The final section discusses the responsibility for resource use and the value of decoupling in making societies compatible with a functional ecosphere.

6.2. Global use of resources

When assessing a country's use of raw materials, we first consider its domestic extractions (DEs), its imports and its exports. This evaluation, as well as the flows of discarded materials, is done using a so-called material flow accounting (MFA) analysis (Krausmann *et al.* 2017a). To understand the functioning of a society as a whole, it is necessary to go beyond the incoming and outgoing flows and know the quantities of materials used in a dissipative way (energy and fertilizers), those used for the constitution of durable goods and infrastructures and, of these, the fraction that is recycled, down-cycled or discarded (waste). "Socio-economic metabolism" refers to the totality of a society's material flows (Figure 6.1).

As can be seen in Figure 6.1, the global extractions (the sum of DEs) are 78 Gt/year (in 2010). One part, 45% in 2010, is intended for dissipative use as an energy source or for food and animal feed (Krausmann *et al.* 2017b). The largest fraction, 55% in 2010, meets the material needs of societies, provides services to the population and supports the economy (buildings, equipment, transport and communication infrastructure, vehicles, machinery, electrical and electronic appliances, etc.). These durably used materials are called "in-use stocks". The stocks thus accumulated over time (accumulated in-use stocks) constitute "manufactured capital"[2]. It is important to know the latter because it needs to be supplied with energy, to be maintained and renewed, so that it determines the future use of energy and materials in the long term. The study of inflows and outflows related to in-use stocks requires an additional and complementary model to the MFA, for example

2 As opposed to "natural capital", i.e. natural resources (biotic and abiotic) and the services that nature provides us with. Manufactured capital always originates from natural capital.

the material inputs stocks and outputs (MISO) model (Wiedenhofer *et al.* 2019).

The global construction of in-use stocks was fueled in 2010 by a material flow of 36 Gt/year, compared to 1 Gt/year in 1900 (Figures 6.1 and 6.2). Over the period 1900–2010, global material stock accumulation reached 792 Gt (Figures 6.1 and 6.2), or 115 t per capita (115 t/cap) on average (Krausmann *et al.* 2017b). This value depends on the country. While this material accumulation was 335 t/cap in 2010 for rich countries, China's was 136 t/cap in 2010, while for the rest of the world it averaged 38 t/cap (Krausmann *et al.* 2017b).

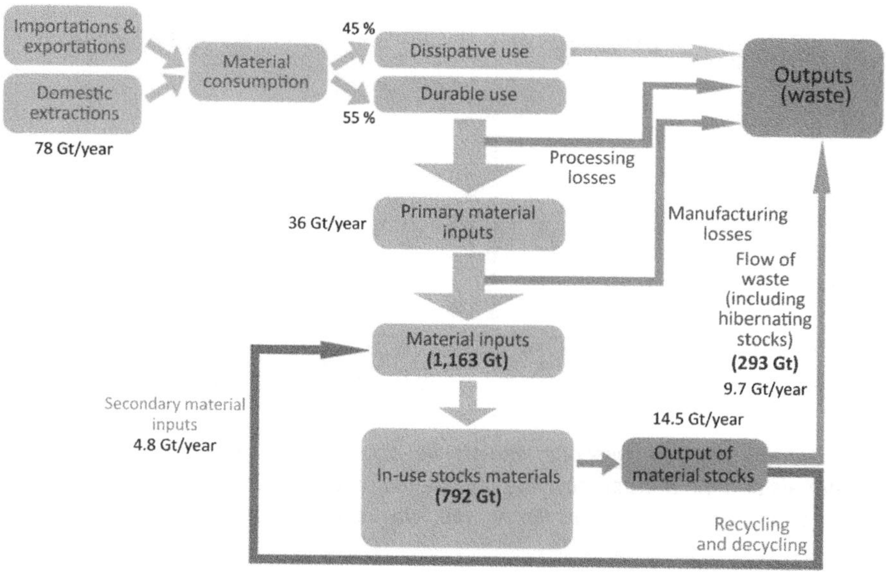

Figure 6.1. *Global balance of material flows (socio-economic metabolism). The numbers in black show the annual flows in 2010, the numbers in purple show the cumulative quantities over the period 1900–2010 (source: adapted from Krausmann* et al. *(2017b)). For a color version of this figure, see www.iste.co.uk/fizaine/mineral2.zip*

In these global approaches, natural resources are generally separated into four categories: biomass (wood and paper), fossil fuels (plastic and asphalt), metals (iron/steel, aluminum, copper, etc.) and non-metallic minerals or more simply "minerals" (concrete, bricks, sand and gravel). Figure 6.2 shows that the accumulated material stocks are dominated by non-metallic minerals, which accounted for 88% of in-use stock inputs in 2010, a value that reflects the intense urbanization of the 20th century (Krausmann *et al.* 2017b). Metals accounted for 3.4% of accumulated materials. Non-renewable resources accounted for a total of 98.3% of the world's manufactured heritage.

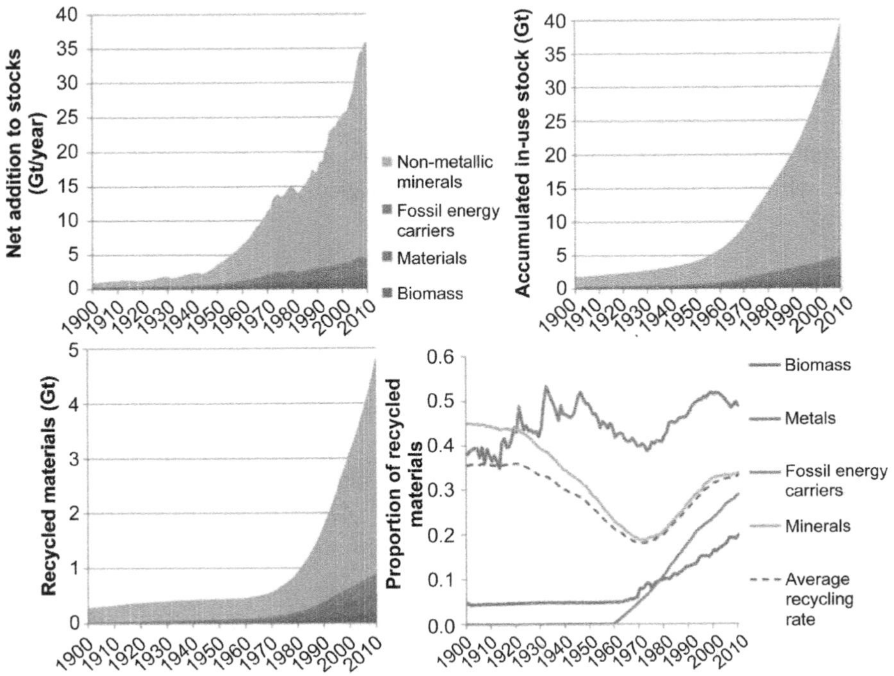

Figure 6.2. *World flows of in-use stocks, accumulated stocks and recycled materials and the annual recycling rate (data source: Krausmann* et al. *(2017b)). For a color version of this figure, see www.iste.co.uk/fizaine/mineral2.zip*

The flow of end-of-life materials from in-use stocks (outputs) increased continuously over the 20th century. In 2010, they amounted to 14.5 Gt (Krausmann *et al.* 2017b). The recycling[3] rates for minerals and metals in 2010 were 32% and 50%, respectively. Thus, secondary mineral inputs and recycled metals accounted for 11% and 22% of total in-use inventory inputs (input recycling rates), respectively (Krausmann *et al.* 2017b). The total recycling rate fell gradually over the century (Figure 6.2) and began to

3 The "output recycling rate", or simply "recycling rate", is the ratio of the secondary input stream, i.e. recycled and decycled materials, to the waste stream (end-of-life outputs). The "input recycling rate" is the ratio of the secondary input stream to the sum of the input streams (primary and secondary inputs).

increase from the 1970s onward, probably due to better regulation and greater technical capacity (Krausmann *et al.* 2017b). By 2010, about one-third of the material was recycled. For the time being, therefore, socio-economic functioning is essentially linear.

6.3. Material consumption indicators

The total amount of materials used globally does not reflect the extent of trade, which represents a significant part of the environmental impact. Indeed, traded goods are responsible for part of the resource use and environmental degradation necessary for their production in producing countries. For example, international trade accounts for ~20% of global freshwater consumption (Hoekstra and Mekonnen 2015), ~25% of total CO_2 emissions (Davis and Caldeira 2010) and ~40% of all materials removed (Wiedmann *et al.* 2015; UNEP 2016). Metal ores are particularly associated with international trade, such as iron (62% of extractions) and bauxite (64%) (Wiedmann *et al.* 2016).

Each country exports part of its resources and imports part of them. To assess material flows, domestic material consumption (DMC) is often used. It takes into account DEs plus imports of physical goods (IM) and subtracts exports of physical goods (EX):

$$\text{DMC} = \text{DE} + \text{IM} - \text{EX}$$

The DMC per capita is also known as the *metabolic rate*. The DMC is an indicator traditionally used at the policy level for national assessments, notably by the OECD and the European Union. It takes into account the physical transfers of products through borders. It does not, however, take into account the auxiliary raw materials that were needed upstream to manufacture consumer products in exporting countries. For example, ores require more material extraction than metals that are actually purchased by importing countries. Similarly, animal or dairy products incorporate more biomass (for animal husbandry) than they physically contain. These quantities of additional materials resulting from the production of commercial products are substantial and can be even greater, sometimes by a factor of 10, than the materials that physically make up the goods being traded.

These "hidden" materials are referred to as *matter embodied in trade* or *indirect flows*. By extension, it is referred to as *impact embodied in trade*, corresponding to the environmental impact generated locally. This idea is well-known for freshwater ("virtual water"; Hoekstra and Mekonnen 2015) and the CO_2 emissions ("carbon leakage"; Davis and Caldeira 2010) that are required for the extraction, manufacture and transportation of manufactured goods. In the case of materials, these are referred to as raw material equivalents of import/export ($RME_{IM/EX}$).

The DMC ignores this indirect impact. It can be considered as underestimating part of the materials consumed by importing countries by attributing the weight of their impacts to exporting countries only. In order to compensate for this shortcoming, another indicator was proposed in 2009 (Muñoz *et al.* 2009; Weinzettel and Kovanda 2009) named raw material consumption (RMC), which accounts for materials incorporated into trade.

This final indicator is referred to as the material footprint (MF):

$$MF = RMC = DE + RME_{IM} - RME_{EX}$$

The MF thus attributes the use of resources needed for extraction and processing to the countries that ultimately consume them, regardless of where they were extracted. The DMC is referred to as a production-based indicator and the MF as a consumption-based indicator (Wiedmann *et al.* 2015; Krausmann *et al.* 2017b). More than the DMC, the MF is seen as a good indicator of the material standard of living (UNEP 2016).

To assess the determinants of material use, the DMC is sometimes broken down into the so-called "IPAT" identity[4] (Ehrlich and Holdren 1971; Blanco *et al.* 2014):

$$DMC = P \times \frac{GDP}{P} \times \frac{DMC}{GDP}$$

where P is population, the second term is GDP per capita or "income" and is an economic indicator and the third is called material intensity (MI). The

4 The IPAT identity is given by: $I = PAT$, where I represents a given environmental impact (resource extraction, pollution flows), P is the population, A is affluence and T is a factor related to the technology used.

latter is a parameter indicating the amount of material that allows a unit of GDP to grow. It can be influenced by economic and technological factors. The inverse quantity, GDP/DMC, is called material productivity (MP) and represents the technical-economic "efficiency" of resource use. Similar quantities can be defined for the MF.

Figure 6.3 shows the DMC and MF per capita for different types of countries: a predominantly exporting country (Chile), predominantly importing countries (France, Japan, the United Kingdom and the United States), and countries with similar exports to their imports (Canada, China and India). The countries that export the most have higher DMCs than their MFs, a situation that is reversed for countries that import the most. Emerging countries (Chile, China and India) see their DMCs and MFs increase rapidly over time due to their rapid socio-economic development. For developed countries (France, Japan, the United Kingdom and the United States), the DMC seems to plateau or slightly decrease. The MF appears to be increasing over time. Despite a major reduction during the economic crisis of 2008, the upward trend resumed thereafter.

The use of materials is particularly high in periods of intense development, as was the case after World War II in the West. As countries "mature" (industrialize), their imports increase relative to exports and their MF becomes dominant over their DMC. Industrialization, or "metabolic transition", is also marked by the conversion from an agrarian economy with low GDP, largely based on biomass and a low environmental impact, to a rich, "mature" industrial socio-economic system with high GDP, high material consumption and a focus on minerals, metals and fossil fuels.

The differences between the DMC and MF highlight the effects of the international market. On a global scale, when considering materials embodied in trade, international trade "weighed" ~30 Gt at the turn of the decade (Wiedmann *et al.* 2015; UNEP 2016), representing ~40% of total annual resource use and ~3 times more than the raw materials that physically cross borders (~10 Gt) (Wiedmann *et al.* 2015; UNEP 2016). Over the entire 20th century, global DMC increased progressively, at a rate below economic growth and above population growth (Krausmann *et al.* 2009). As a result, the MI declined continuously. The annual growth of the MF at the beginning of the 21st century was higher than that of the DMC (4.6% vs. 3.7%, respectively) (UNEP 2016).

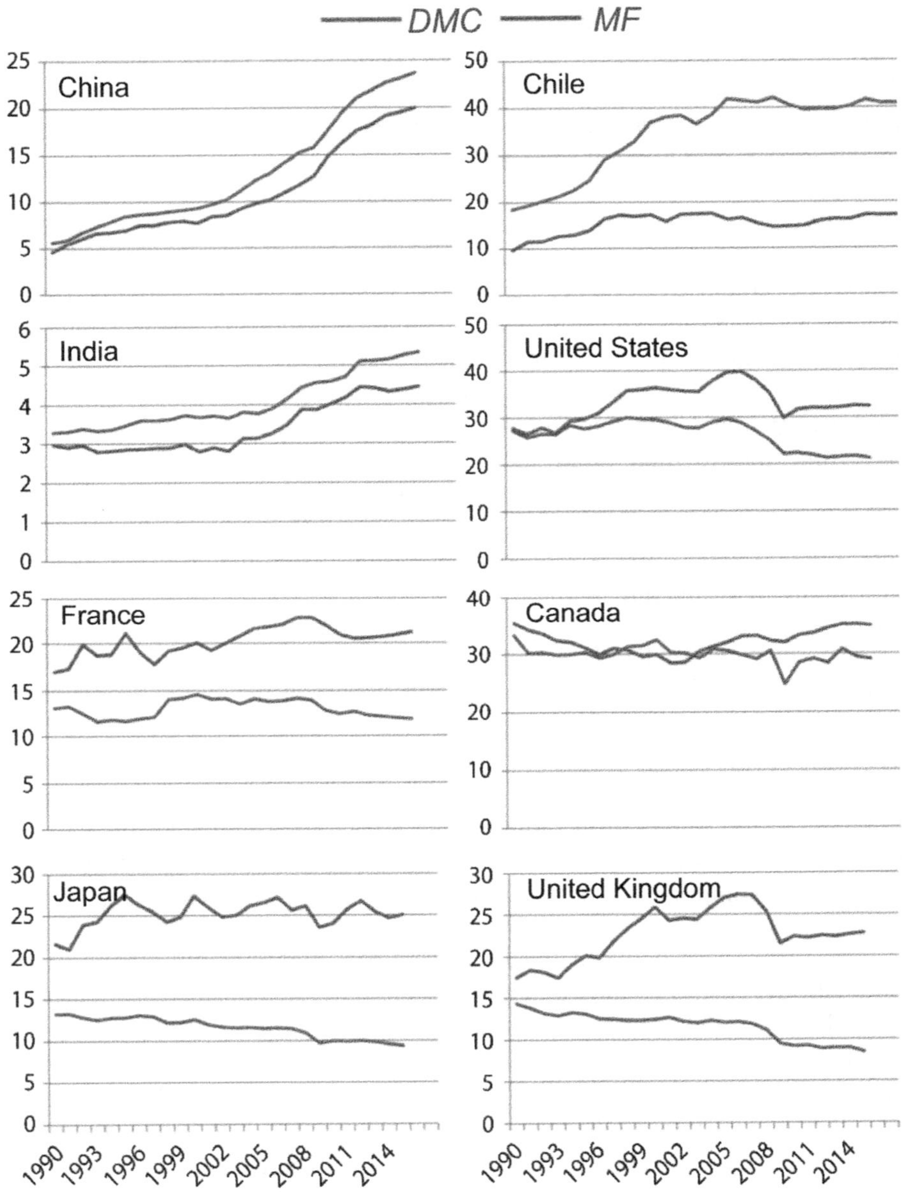

Figure 6.3. *DMC and MF in tons per capita between 1990 and 2015 for different countries. Note that the y-axis scales are different (sources: Wiedmann* et al. *(2015); UN Environment International Resource Panel, Global Material Flows Database). For a color version of this figure, see www.iste.co.uk/fizaine/mineral2.zip*

6.4. Decoupling the economy from resource consumption

6.4.1. *Evidence of decoupling*

As the exploitation of natural resources has steadily increased over time, it has resulted in a strong increase in living standards, health benefits and improved comfort. Unfortunately, this progress has been accompanied by severe environmental degradation. Above all, this mainly benefits the rich countries, those that consume the most. Other countries are following the West's lead and have been since the 1970s and 1980s. While it is commendable that countries are improving their material living standards, the resulting environmental pressure is becoming untenable and is increasing global dysfunction.

In order to break the vicious circle that links material resources → economy → welfare/environmental degradation, the concept of decoupling has been put forward, in particular by the UNEP (UNEP 2011). The principle consists in decoupling, or dissociating, the use of resources from economic growth (Figure 6.4). The objective is to use fewer materials, thus reducing environmental impact, while increasing the wealth of individuals. The indicator used for wealth is GDP per capita. Decoupling is *relative* if resource use increases at a slower rate than GDP (Figure 6.4); it is *absolute* when the former decreases and the latter increases. Decoupling is a general concept that can target the links between the economy, or another indicator such as human well-being, and the causes of an environmental problem (resource use, CO_2 emissions, etc.) or an indicator of that environmental problem (global warming, deforestation, etc.). Moreover, it can be applied to an economic sector, a state or even the world.

Since MI (material intensity) is given by the ratio of DMC to GDP, it has traditionally been preferred as an indicator of decoupling. A decrease in MI (or an increase in MP) was observed worldwide over the 20th century (Krausmann *et al.* 2009). Fewer materials allowed GDP to grow, which is similar to decoupling. Indeed, absolute decoupling occurs between DMC and GDP for some developed countries, as illustrated by the decrease in DMC in Figure 6.3 for Japan and the United Kingdom. For Canada, the United States and France, DMC appears to be rather constant. These trends are confirmed in Figure 6.5, which compares on the same scale the relative change in DMC, MF and GDP from 1990 over time. The magnitudes of the respective changes in this figure are then comparable.

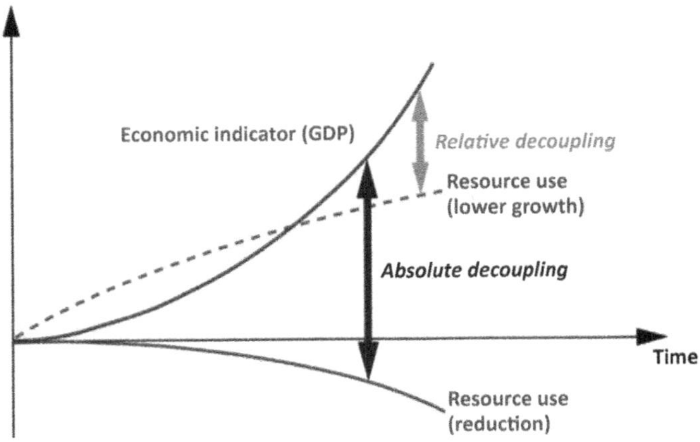

Figure 6.4. *Relative and absolute decoupling of resource use from GDP . For a color version of this figure, see www.iste.co.uk/fizaine/mineral2.zip*

While the DMC of developed countries is decreasing or plateauing as their GDP is increasing, thus tending to dissociate, their MF is increasing, and this at a similar rate to that of GDP. Thus, according to this criterion, decoupling does not take place (Wiedmann *et al.* 2015). This result is attributed to the transfer by developed countries of a significant part of their resource consumption, and the environmental impact that comes with it, to exporting countries (emerging and developing countries) via their imports. Thus, a decrease in MI indicates progress in the eco-efficiency with which developed countries use resources, but not an absolute reduction in their use of materials.

For India and China, the DMC and MF standardized in Figure 6.5 increase in unison, but at a lower rate than their GDP (relative decoupling). For Chile, the growth of the two indicators is similar and close to GDP, although relative decoupling from the MF seems to have been achieved in the early 2000s.

Looking globally in time and across the world, evidence of absolute decoupling of the MF is almost non-existent (Steinberg *et al.* 2013; Wiedmann *et al.* 2015; Pothen and Welsch 2019). The only reported case is South Africa (Wiedmann *et al.* 2015). The distinction between decoupling of resources according to DMC or MF is a very topical issue. The decouplings evaluated in relation to DMC or MF (or RMC) are named *net* and *gross* decoupling, respectively (Laurent 2011).

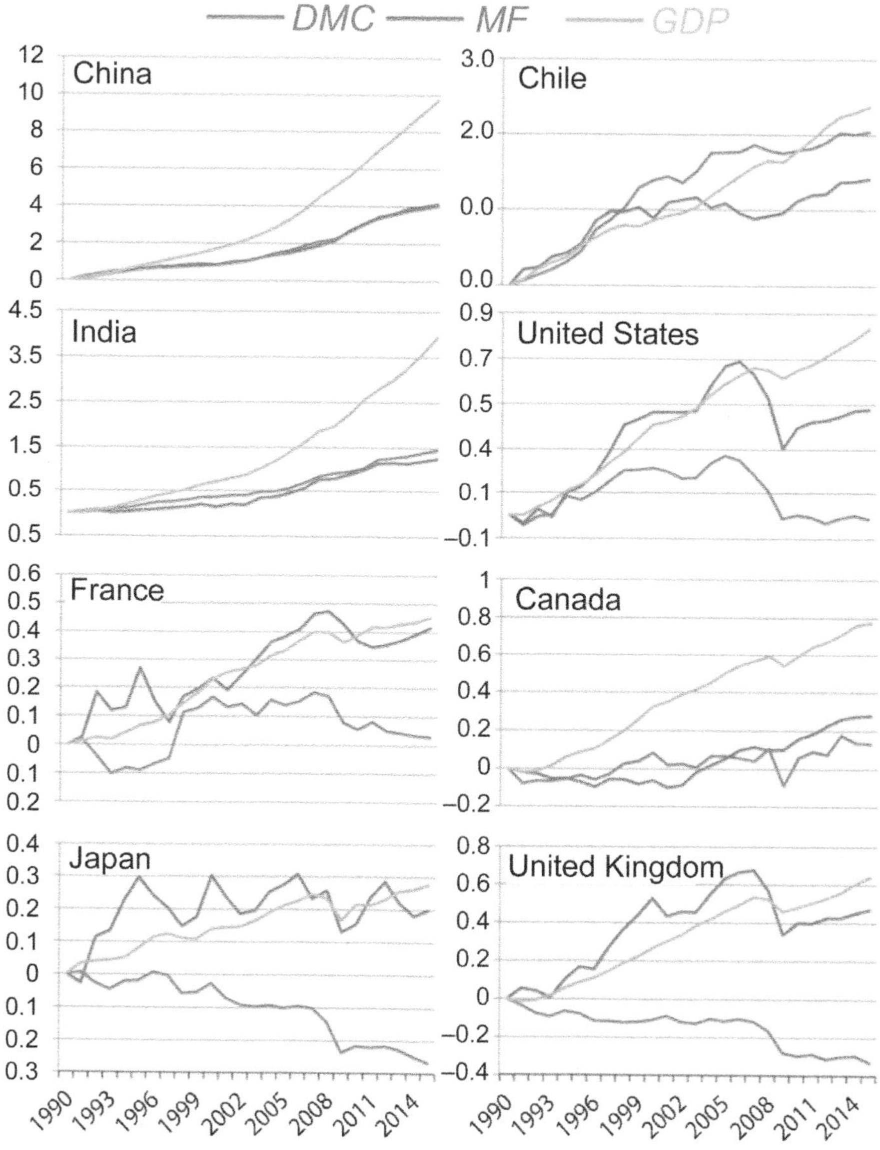

Figure 6.5. *Relative change in DMC, MF and GDP between 1990 and 2015 for different countries. Relative changes are calculated according to x(t) = (x(t) – x(1990))/x(1990) (sources: Wiedmann* et al. *(2015); UN Environment International Resource Panel, Global Material Flows Database). For a color version of this figure, see www.iste.co.uk/fizaine/mineral2.zip*

Analysis of data to quantify the effect of divers of resource use[5] shows that material flows are driven primarily by GDP and to a lesser extent by population size (Steinberg *et al.* 2010; UNEP 2016). Material intensity declined over the 20th century and had a mitigating effect on the impact of wealth and demography. However, while changes in MI were mainly driven by population and income between 1990 and 2000, income became the main driver between 2001 and 2010 (UNEP 2016). Material intensity amplified the effect of population and GDP during this decade (UNEP 2016). This observation may have implications for assessing the links between resource use and income.

Analyses by material type show that the use of ores, industrial minerals and fossil fuels is closely coupled with income (Steinberg *et al.* 2010, 2013; Plank *et al.* 2018; Pothen and Welsch 2019). These materials become predominant in countries that have undergone a metabolic transition. They are thus very unequally shared across the world, with distribution according to economic differences (Steinberg *et al.* 2010). The use of biomass is moderately related to economy and is mainly dictated by population (Krausmann *et al.* 2009; Steinberg *et al.* 2010, 2013; Pothen and Welsch 2019). Biomass is preponderant in non-industrialized countries. It is less unevenly distributed than non-renewable materials so it is at the margin of the other resources. In particular, it appears to be an essential resource for livelihoods (Krausmann *et al.* 2009; Steinberg *et al.* 2010).

Fewer studies have covered analysis of the factors that govern the growth of MF. Changes in the global MF seem to be strongly coupled to changes in GDP (Wiedmann *et al.* 2015; Pothen and Welsch 2019). As for the DMC, the MF of biomass is less correlated with GDP, whereas non-renewable resources are strongly related to it, in particular fossil fuels, which show the strongest dependence (Wiedmann *et al.* 2015; Pothen and Welsch 2019).

6.4.2. *Saturation of resource use*

It is expected that countries will ultimately see a decline in consumption as they improve MI as well as having less need to build infrastructure and seeing their consumption of goods eventually saturate. Thus, the change in environmental impact as a function of GDP should follow an inverted

5 This is a multivariate regression decomposition according to the IPAT equation described above. The details of the method are beyond the scope of this chapter and the interested reader will find details in Steinberg *et al.* (2010) and Wiedmann *et al.* (2015).

U-curve or bell curve, a hypothesis called the environmental Kuznets curve (EKC). The first part of the curve corresponds to developing societies marked by a strong increase in the use of resources and a large material footprint; the second part corresponds to developed societies that, *a priori*, implement better regulations and develop more efficient technical means.

In a study of the United States over more than a century, it was found that various products, or classes of products, have reached or are on the verge of reaching a limit, suggesting saturation of consumption (Chen and Graedel 2015). At the global level, however, there is no indication of resource consumption saturation, even for developed countries (Krausmann *et al.* 2017b). Beyond the growing demand and the mechanisms that link resource use and economic growth, the expansion of international trade is also increasing its contribution. One reason is that developed countries are increasing their imports from countries with lower material productivity (Plank 2018).

However, one study suggests that saturation of material demand has already been achieved by some industrialized countries for certain categories of materials (Bleischwitz *et al.* 2018). This result is consistent with the plateau or decrease in DMC noted above. According to this study, China is on the way to saturating the use of certain materials. Since saturation may require a delay of 30 years or more between the use of a material and the leveling off of its growth, this work suggests that emerging countries will experience material saturation in the decades to come (Bleischwitz *et al.* 2018).

Other authors have confirmed that in the long term, developing and emerging countries experience a close coupling between DMC and GDP for all categories of materials (Steinberg *et al.* 2013). Developed countries have a lower coupling. Evaluating different models, Pothen and Welsch (2019) showed that the EKC hypothesis could not adequately represent the GDP distribution of all countries. The hypothesis only works if OECD countries alone are considered. An S (or sigmoid) curve better represents the use of resources with the global distribution of GDP. However, the regression remains imperfect and saturation is only reached for a GDP well above the maximum actually observed. A linear relationship seems to better represent the data with proportionality factors of 0.562 and 0.752 for the DMC and the MF, respectively (Pothen and Welsch 2019). The latter value is similar to the value (0.60) found by Wiedmann *et al.* (2015). Thus, DMC and MF seem to increase less rapidly than wealth, and MF is more sensitive to economic growth (Pothen and Welsch 2019).

The evidence of absolute decoupling for developed countries in the case of DMC is encouraging, but its absence for MF is worrisome, although some models suggest that it might occur. The fact that emerging countries adopt the economic paradigm of developed countries suggests a large future ecological footprint. The saturation of consumption from a certain income level is neither demonstrated nor seems likely to be sufficient to reduce the MF below a threshold compatible with planetary boundaries. The mitigating effects of the MP are too tenuous to significantly curb material growth as it is linked to economic growth and population. The efficiency gains thanks to technology, although they can contribute positively to the reduction of MF, appear more often than not to halt dematerialization. The stumbling blocks are above all the rebound effect[6] and the fact that innovation contributes endogenously to the economy and economic growth (Laurent 2011; Steinberg *et al.* 2013; Sachs 2015; Lefèvre 2016). The strong GDP resources dependence thus leaves one wondering about the possibility of decoupling in the current system. It even seems that dematerialization implies lower growth (Steinberg *et al.* 2013).

The scientific data thus lead to the following observation:

> [...] many scholars doubt that absolute decoupling of economic growth from material use and environmental impact is possible at all and instead call for a far-reaching transformation of the economic system and a shift in focus from economic growth toward human well-being or prosperity. (Krausmann *et al.* 2017a)

6.5. Responsibility for resource consumption

It goes without saying that the rise in the standard of living in emerging and developing countries is more than legitimate, but it cannot be achieved if the state of the planet deteriorates excessively. Although very unequally shared, natural resources are less so than GDP, energy or territories (Steinberg *et al.* 2010). The issue of resource use is nevertheless particularly critical. Its importance is accentuated when considering non-renewable resources since these are limited and their overuse undermines the capacity of future generations to live as well as previous generations or to have as

6 The rebound effect, or Jevons paradox, states that, contrary to what is expected, improved energy or material efficiency leads to an increase in consumption.

many material choices for their development. In the shorter term, there is a need to reduce material consumption simply because the quantities currently used and their growth are such that their consequences may soon exceed an acceptable material and civilizational footprint. At the global level, absolute decoupling is unavoidable. In this context, and for the sake of equity, it is obvious that the countries that consume the most materials will have to use less to allow others to meet their needs.

Developed countries use the bulk of the annual extractions of materials. The richest 10% use 27% of the materials consumed annually (Steinberg *et al.* 2010). As discussed above, consumption in developing countries is even higher when considering market-related material flows. While the MF for developed countries is 20–25 t/cap, it is 4 t/cap for the poorest countries (Wiedmann *et al.* 2015; UNEP 2016). In terms of DMC, the OECD consumes 28% of the world's resources, but in terms of MF, this number rises to 38% (UNEP 2016).

These two indicators, DMC and MF, do not take the same view. DMC places the responsibility for resource use on producer (exporting) countries, while in the latter, this responsibility falls on consumer (importing) countries. The DMC underestimates resource use for importers, so the MF could be the preferred indicator at the policy level to assess material impact. This argument is perfectly justified since importing countries are responsible for their consumption and imports (RME_{IM}), with the effect of transferring the environmental impact of resource extraction and processing to the exporting countries. However, assessing resource use according to the MF releases producers from their own responsibility. The point of view can therefore also be reversed: the MF underestimates the resource withdrawals of exporters, whereas the latter benefits economically from their exports.

The question ultimately comes down to who is responsible for the extraction and transformation of natural resources in a globalized trading system: the producers or the consumers? Underlying this is the question of who is responsible for the over-exploitation of resources by civilization, producer countries or consumer countries? A similar question arises more broadly: who is responsible for the impact of over-consumption? The customer, the manufacturer or the seller? It turns out that none of them can *a priori* avoid their responsibilities, neither at the local nor at the international level. All are responsible, because they are all cogs that make the system work, influence it and benefit from it. The two decoupling indicators therefore appear useful and complementary, particularly for international policies. They allow a better evaluation of the role of each. From this perspective, these indicators are potentially useful for setting relevant quantitative targets. The MF, for example, gives a more realistic picture of material consumption, especially for rich countries.

Developed countries have indeed a particular historical responsibility for the global environmental state. Their development spans more than 250 years and they have experienced considerable resource use, which has led to many of today's environmental problems. This observation is well known in the negotiations of the United Nations Framework Convention on Climate Change under the term "common but differentiated responsibility". The same notion obviously holds for natural resources. Thus, rather than considering annual flows, it may be preferable to account for the cumulative sum of activities responsible for environmental degradation (use of natural resources, water, territories, CO_2 emissions, etc.). This is particularly important because emerging countries, notably China, are catching up with developed countries in their environmental impact, so their responsibility is increasing.

A recent study has taken a historical approach by assessing the cumulative use of resources around the world between 1950 and 2010 (Mayer *et al.* 2017). Looking at the year 2010, industrialized countries (Western Europe and North America) accounted for 12% of the world's population and 16% of the world's DEs. Over the period 1950–2010, industrial countries accounted for 15% of the population but used 29% of the world's DEs. Cumulative imports from these countries amounted to 59% of the world total between 1950 and 2010 but 44% for the year 2010. The picture is reversed for China (Mayer *et al.* 2017). In 2010, China accounted for 20% of the world's population and took 32% of the world's resources. The cumulative values between 1950 and 2010 were 22% and 17%, respectively. China's imports decreased from 15% of the world market in 2010 to 5% for the cumulative value. Thus, while, in a historical perspective, developed countries have a resource use double that of 2010, China's resource use has been halved (Mayer *et al.* 2017). This work suggests that it would be relevant for international agreements to take this reality into account in setting their targets.

6.6. Conclusion

A minimum level of resources is necessary for human well-being. In general, consumption increases with wealth. However, the global picture is more complex, because for the same income, consumption can vary by a factor of nine, while for a given resource consumption, income can vary by a factor of 20 (UNEP 2016). It is therefore possible to live with less for the same level of wealth.

Nevertheless, the overall trends are worrying. As described above, the material decoupling anticipated for rich countries has not been achieved. On the contrary, material consumption appears to be totally linked to economic growth and there is no income at which material use ceases (UNEP 2016). How then can we reconcile the development of developing and emerging countries and the maintenance of well-being in developed countries, while reducing the global material footprint and respecting planetary boundaries?

Decoupling was proposed as an objective to be achieved. Very often, it is desired to promote equitable development of countries. Decoupling is thus associated with sustainable development and green growth or the green economy, i.e. a context of continuous economic growth. However, there is no evidence of the future success of this vision.

The picture is overshadowed by the fact that, with development countries are using more and more non-renewable materials – metals, minerals and fossil fuels – which can ultimately lead to shortages. In the event of scarcity, supply difficulties could arise for some countries for economic and geopolitical reasons. In order to make societies sustainable, renewable resources should be favored, although care should be taken not to withdraw them at a higher rate than their regeneration or to withdraw them at the expense of land intended for food and the preservation of ecosystems.

There is a great deal of reliance on improving MI through technological innovation. However, current evidence does not support the idea that advances in energy or material efficiency can counteract the rate of increase in global material inputs. Nor is there any evidence that technology can substitute for degraded ecosystems and compensate for the loss of the services they provide to humans. More efficient technology can even make the situation worse because of the rebound effect.

One way to reduce the extractions would be to move from a linear economy, where the recycling rate is low, to a circulating economy, where the recycling rate would significantly reduce the use of new resources. Reducing material consumption requires that material flows be part of a closed cycle. There are, however, fundamental limits to this principle. In addition to the energy required and the losses during recycling processes, the input of recycled materials (secondary inputs) cannot compensate for the current growth rate of primary inputs (Arnsperger and Bourg 2016). Circularity must absolutely be accompanied by a reduction in inputs. Thus, recent projections on global societal metabolism show that a reduction in primary inputs of stocks in use by 90% between 2015 and 2030, combined with a 90% increase in the recycling rate, will lead in 2050 to a reduction in the quantity of waste to only 24% of the initial rate (2.8 Gt/year) (Wiedenhofer *et al.* 2019). Moreover, under this scenario, the accumulated in-use material stocks will still increase in 2050 by 42% of the initial value (1,314 Gt).

A truly circular economy, more energy and material efficient products, and more sustainable products are obviously essential to dematerialize the socio-economic system, but they are just as obviously insufficient. Societies must also and above all reduce primary resource inputs in an absolute way. Basically, absolute decoupling must be implemented. However, because of the unequal sharing of resources, wealth and well-being, the responsibility for the transformations necessary for decoupling must fall primarily on the developed countries.

The current limits of decoupling may be due to the fact that the quantities of resources are compared to GDP, a purely economic indicator that does not necessarily reflect the well-being or happiness of populations. If the use of resources and the associated impacts increase with income, a decoupling indicator that includes social and environmental criteria is more relevant. It is essential to understand whether a certain quantity of resources is sufficient to ensure a high quality of life and, if so, under what conditions. Several quality-of-life indices have been developed. For example, increased resource use is correlated with a higher value of the "social progress index" over the period 1950–2015 (Mayer *et al.* 2017). Disparities between countries are very strong according to this index and a higher quality of life requires a minimum of 460 t/cap of accumulated material. The amount of resources also increases with the Human Development Index (HDI), but less and less rapidly as the HDI increases (UNEP 2016). The effect can be observed for both the DMC and the MF. Moreover, the consumption of resources required for a high HDI appears to decrease over time. It is mainly the educational and life expectancy aspects of the HDI that seem to benefit, not income. This effect is interpreted as reflecting an accumulation of materials in infrastructure that has long-term effects on education and public health, while income is more dependent on material inputs (UNEP 2016).

Current trends in the consumption of natural resources and the weight of the economy show that absolute decoupling is currently out of reach. The positive effects of technological progress, recycling and the maturation of societies as advocated by the proponents of green economies do not allow us to understand the close coupling between materials and GDP. An absolute decoupling of resources is nevertheless a laudable and unavoidable objective in order to make societies sustainable. However, this target necessarily involves reducing inputs and an accompanying high recycling rate. It is also necessary to redefine both the indicator for measuring decoupling and the plans to achieve the desired objective. According to scientists' own observations, decoupling will not be possible in the current socio-economic system.

6.7. References

Arnsperger, C. and Bourg, D. (2016). Vers une économie authentiquement circulaire – réflexions sur les fondements d'un indicateur de circularité. *Revue de l'Observatoire français des conjonctures économiques*, 145, 235–257.

Blanco, G., Gerlagh, R., Suh, S. (2014). Drivers, trends and mitigation. In *Climate Change 2014: Mitigation of Climate Change. Contribution of Working Group III of the Intergovernmental Panel on Climate Change to the 5th Synthesis Report*, Edenhofer, O. *et al.* (eds). Cambridge University Press, New York.

Bleischwitz, R., Nechifor, V., Winning, M., Huang, B., Geng, Y. (2018). Extrapolation or saturation – Revisiting growth patterns, development stages and decoupling. *Global Environmental Change*, 48, 86–96.

Chen, W.-Q. and Graedel, T.E. (2015). In-use product stocks link manufactured capital to natural capital. *Proceedings of the National Academy of Sciences of the USA*, 112, 6265–6270.

Davis, S.J. and Caldeira, K. (2010). Consumption-based accounting of CO_2 emissions. *Proceedings of the National Academy of Sciences*, 107, 5687–5692.

Dittrich, M., Giljum, S., Lutter, S., Polzin, C. (2012). Green economies around the world? Implications of resource use for development and the environment [Online]. Available at: https://www.boell.de/sites/default/files/201207_green_economies_around_the_world.pdf [Accessed April 2019].

Ehrlich, P.R. and Holdren, J.P. (1971). Impact of population growth. *Science*, 171, 1212–1217.

Hoekstra, A. and Mekonnen, M. (2015). The water footprint of humanity. *Proceedings of the National Academy of Sciences*, 109, 3232–3237.

Hoekstra, A. and Wiedmann, T. (2014). Humanity's unsustainable environmental foot-print. *Science*, 338, 1114–1117.

Krausmann, F., Gingrich, S., Eisenmenger, N., Erb, K.-H., Haberl, H., Fischer-Kowalski, M. (2009). Growth in global materials use, GDP and population during the 20th century. *Ecological Economics*, 68, 2696–2705.

Krausmann, F., Schandl, H., Eisenmenger, N. *et al.* (2017a). Material flow accounting: Measuring global material use for sustainable development. *Annual Reviews of Environment and Resources*, 42, 647–675.

Krausmann, F., Wiedenhofer, D., Lauk, C., Haas, W., Tanikawa, H., Fishman, T., Miatto, A., Schandl, H., Haberl, H. (2017b). Global socioeconomic material stocks rise 23-fold over the 20th century and require half of annual resource use. *Proceedings of the National Academy of Sciences*, 114, 1880–1885.

Krausmann, F., Lauk, C., Haas, W., Wiedenhofer, D. (2018). From resource extraction to outflows of wastes and emissions: The socioeconomic metabolism of the global economy, 1900–2015. *Global Environmental Change*, 52, 131–140.

Laurent, E. (2011). Faut-il décourager le découplage ? *Revue de l'Observatoire français des conjonctures économiques*, 120, 235–257.

Lefèvre, T. (2016). *Sortir de l'impasse – Qu'est-ce qui freine la transition écologique ?*. Multimondes, Montreal.

Mackenzie, F.T. (2011). *Our Changing Planet – An Introduction to Earth System Science and Global Environmental Change*, 4th edition. Prentice Hall, Boston.

Mayer, A., Haas, W., Wiedenhofer, D. (2017). How countries' resource use history matters for human well-being – An investigation of global patterns in cumulative material flows from 1950 to 2010. *Ecological Economics*, 134, 1–10.

Muñoz, F., Giljum, S., Roca, J. (2009). The raw material equivalents of international trade empirical evidence for Latin America. *Journal of Industrial Ecology*, 13, 881–897.

Plank, B., Eisenmenger, N., Schaffartzik, A., Wiedenhofer, D. (2018). International trade drives global resource use: A structural decomposition analysis of raw material consumption from 1990–2010. *Environmental Science and Technology*, 52, 4190–4198.

Pothen, F. and Welsch, H. (2019). Economic development and material use. Evidence from international panel data. *World Development*, 115, 107–119.

Ripple, W.J., Wolf, C., Newsome, T.M., Galetti, M., Alamgir, M., Crist, E., Mahmoud, M.I., Laurance, W.F. (2017). World scientists' warning to humanity: A second notice. *BioScience*, 67, 1026–1028.

Rockström, J., Steffen, W., Noone, K., Persson, Å., Chapin III, F.S., Lambin, E., Lenton, T.M., Scheffer, M., Folke, C., Joachim Schellnhuber, H. *et al.* (2009). Planetary boundaries: Exploring the safe operating space for humanity. *Ecology and Society*, 14, 32.

Sachs, J. (2015). *The Age of Sustainable Development*. Columbia University Press, New York.

Steffen, W., Richardson, K., Rockström, J., Cornell, S.E., Fetzer, I., Bennett, E.M., Biggs, R., Carpenter, S.R., de Dries, W., de Wit, C.A. *et al.* (2015). Planetary boundaries: Guiding human development on a changing planet. *Science*, 347, 1259855.

Steinberg, J.K., Krausmann, F., Eisenmenger, N. (2010). Global patterns of materials use: A socioeconomic and geophysical analysis. *Ecological Economics*, 69, 1148–1158.

Steinberg, J.K., Krausmann, F., Getzner, M., Schandl, H., West, J. (2013). Development and dematerialization: An international study. *Plos One*, 8, e70385.

UNEP (2011). L'humanité peut et doit faire plus avec moins, souligne le PNUE. International Resource Panel, Nairobi.

UNEP (2016). Global material flows and resource productivity – An assessment study of the UNEP International Resource Panel. International Resource Panel, Paris.

UNEP (2019). *Healthy Planet, Healthy People – Global Environment Outlook GEO-6*. Cambridge University Press, Cambridge.

Weinzettel, J. and Kovanda, J. (2009). Assessing socioeconomic metabolism through hybrid life cycle assessment – The case of the Czech Republic. *Journal of Industrial Ecology*, 13, 607–621.

Wiedenhofer, D., Fishman, T., Lauk, C., Haas, W., Krausmann, F. (2019). Integrating material stock dynamics into economy-wide material flow accounting: Concepts, modelling, and global application for 1900–2050. *Ecological Economics*, 156, 121–133.

Wiedmann, T.O. (2016). Impacts embodied in global trade flows. In *Taking Stock of Industrial Ecology*, Clift, R., Druckman, A. (eds). Springer, Heidelberg.

Wiedmann, T.O., Schandlb, H., Lenzen, M., Moran, D., Suh, S., West, J., Kanemoto, K. (2015). The material footprint of nations. *Proceedings of the National Academy of Sciences*, 112, 6271–6276.

7

The Economics of Recycling: Ambitions, Myths and Constraints

Alain GELDRON
Agence de l'environnement et de la maîtrise de l'énergie (ADEME), Angers, France

The recycling of mineral raw materials concerns a wide range of issues. It is especially developed for metals and glass, the recycling of which is part of the same dynamic. However, it is important not to forget aggregates, which represent a much larger tonnage but whose markets, technologies and constraints are of different natures. In the same way, industrial minerals, refractories, abrasives and others can be recycled in different material contexts. In this chapter, we will deal more specifically with metals.

7.1. The recycling economy, an ancient history

Although in some minds recycling started in the early 1990s with the launch of selective collection of household packaging, it actually dates back far into prehistory. As early as the Paleolithic period, more than 500,000 years ago, there is evidence of the recycling or reuse of bone or stone tools (Amick 2014), particularly in Tabun Cave in Israel (Shimelmitz 2015). This recycling led to the object being reworked and therefore to a reduction in the material actually used.

Mineral Resources Economics 2,
coordinated by Florian FIZAINE and Xavier GALIÈGUE.

With the appearance of the use of metals during the Bronze Age, 5,000 years ago, the economy of recycling seemed to develop through exchanges as shown by the used French bronze objects transported to Great Britain and discovered on the coast of Dover (Kuijper 2019). Recycling continued to develop afterward with the use of iron. During the Greek and Roman empires, workers preferred to remelt metals already produced (Béguin 2013). Although there is some work on the origin and evolution of metallurgy (Center for Archaelogy Guidelines 2001), there is little literature on the place of metal recycling in the economy until recently. Various papers find the origin of a recycling process structuring a sector in the 1030s with the development of paper recycling in Japan. The story then becomes that of urban waste, "garbage", in connection with insalubrity until the end of the 19th century, which saw the golden age of ragpickers whose activity focused on materials of organic origin (Barles 2005). With the advent of industrial society, which saw the development of the use of metals in the first half of the 20th century, the scrap metal trade developed.

Particularly developed during periods of shortages and especially armed conflicts, metal recycling is becoming a component of the raw materials economy. It there constitutes either a craft activity or a means of subsistence for marginalized populations. Until the beginning of the 1990s, recycling was established as an activity of recuperators from family businesses specialized in certain material flows.

In 1992, Germany and France introduced extended producer responsibility for used packaging. Subsequently studied and conceptualized by the work of the OECD (2001)[1], it aimed, on the one hand, to shift the financing of the management of this invasive waste from local authorities to producers (manufacturers, importers and distributors) and, on the other hand, to encourage the latter to take preventive measures. It also changed the economic balance of recycling by making it the responsibility of the companies that put the products on the market. This is a form of the "polluter pays" principle enshrined in the European directive on waste and in the 1975 French law on waste.

The 3Rs (Reduce, Reuse, Recycle) initiative taken by Japan and concretized by the April 2005 ministerial meeting in Tokyo (Japanese Ministry of the Environment 2005) reinforced the place of recycling in the policies of various countries as a key element in waste management. While recycling was part

1 This manual has been updated (OECD 2017).

of the model of a somewhat marginal economy, it was reinforced by a new impetus, one of an environmental nature. While the new and complementary interest in having access to a resource for industry was taken into account in Japan in the early 2000s, it was not until the beginning of 2010 that this dimension was introduced in Europe, and particularly in France, into public policies.

Since the first oil shock, recovery companies have been undergoing profound changes. There has been a decrease in the number of companies and an increase in their sizes, as well as the development of downstream industrial companies for final treatment. The industrialization of the sector has been accompanied by a convergence between the traditional stakeholders in the waste sector, who are involved in landfilling and incineration, and those in recovery, as well as by a greater mix of activities. At the same time, with the development of extended producer responsibility policies, particularly in Europe, recycling has gradually expanded from single-material waste (such as glass bottles), whether from production or end-of-life products, to complex waste, such as used electrical and electronic products.

Recycling has gone from being a marginal issue to an area of the future mobilizing multiple challenges, including the development of local economic activity while at the same time limiting environmental impact, particularly in the fight against climate change. It can also reduce external dependence and the exploitation of virgin resources.

7.2. Geological and urban mines, similarities and differences in logic

As in the field of geological mining, the field of urban mining[2], as it has often been called since the early 2000s, is based on the notion of a deposit applied to waste. The chain of mobilization of the materials of both fields presents some similarities, with a first stage of extraction of minerals on the one hand and collection on the other hand, a second stage of concentration/ preparation from the ore or waste and finally a third stage of a pyro- or hydro-metallurgical nature. The treatment of waste, particularly metallic waste, is largely inspired in its principles and tools by mineral and metallurgical processes. However, as recycling progresses, the tools and methods are becoming more and more specific and adapted to the characteristics of waste.

2 The trademarks of the words "*mines urbaine*" (urban mine) and "*mines urbaines*" (urban mines) are respectively the property of RECUPYL and Fondation Paris Tech.

However, mining and recycling activities differ significantly. First, natural deposits are localized concentrations of ores that can contain several tens of millions of tons, for base metals or a few tons for precious metals, over a limited area of a few square kilometers, whereas for waste, there are highly dispersed deposits that must be gathered together to be transported to treatment facilities. The logistics and the economy are thus totally different, with, on the one hand, often multinational capital companies and, on the other hand, most often small to medium structures with more limited investment and possibly, depending on the country and the period, a significant part in the informal economy.

The second difference is the nature of the materials to be treated. Waste, unlike minerals, contains carbon chains, mainly plastic, and also coatings, such as paints or varnishes. These carbonaceous contents disturb certain treatments, such as the pyrometallurgical treatment of electronic cards, by generating too much gas and heat for the proper management of the furnace.

	Average grade in ore expressed in metal unless otherwise stated	Average content in waste expressed as metal	Waste concerned
Gold	1–10 g/t	1 to more than 200 g/t	Electronic cards
Platinoids	1–8 g/t	260 g/t	Catalytic converters
		100 g/t	Electronic cards
Rare earths	1–15% (expressed in oxides)	35%	Permanent magnets
		5,000 g/t	Compact fluorescent lamps
Indium	1–100 g/t	175 g/t	LCD screens
Cobalt	0.3%	15%	Li-ion batteries
Tantalum	250 g/t Ta_2O_3	3,000 g/t	Mobile phones
Lithium	0.04% for salt and 1% for mines	1.5%	Li-ion batteries

Table 7.1. *Comparison of contents in natural deposits and in selected wastes (source: Geldron (2016))*

The third important difference is the diversity of the metals contained and their concentration. In natural deposits, the metals contained and exploited can be limited in number, from a single one for iron ore to three or four or, at most, about ten. In the case of waste, the number of metals in electronic products usually exceeds 40. In waste, the grades are also extremely variable. From near-pure metal or alloy scraps to base metal fabrication and consumption waste in structures, these materials can become very diluted in equipment. However, many rare metals available in nature are more concentrated in waste (Table 7.1).

In some electronic cards, the gold content can even reach more than 3 kilograms per ton, 1,500 times more than in most deposits currently being mined.

This situation leads to the fact that, generally speaking, only those metals are recycled that are found in greater concentration in waste than in ores, due to obvious constraints of economic equilibrium.

The last fundamental difference between the two areas is that the exploitation of minerals represents stock management, whereas waste management corresponds to flow management.

7.3. Understand the definitions and indicators of recycling in order to express its performance

Waste can be classified into three types according to its origin in the lifecycle of the material. At the beginning of the cycle, home scraps with non-conforming characteristics appear, but they are generally recycled directly in production, for example in the steel mill. Then, the process of transforming the material and manufacturing the products generates new scraps and then end-of-life waste (old scraps).

If the waste returns after recycling in the manufacture of the same products, it is called closed-loop recycling; if it returns to another manufacture, usually of lesser value, it is called open-loop recycling.

The literature also distinguishes between overcycling (upcycling) and undercycling (downcycling), which is more applicable to materials. These consist respectively of recycling materials to produce an object of higher value, for example a piece of furniture made from wooden pallets, or in recycling at a lower value, for example the use of glass as a road underlay.

When talking about recycling, it is unavoidable to talk about the recycling rate, but this varies in its content depending on the author. Some studies have formalized the calculation of different rates applying to the field of recycling (Graedel *et al.* 2011). These have been taken up by the work of the International Resource Panel (UNEP IRP 2011) and EUROSTAT (EUROSTAT 2019).

Two common points of confusion in the expression of rates are:

– the recycling rate is confused with the waste collection rate or with the efficiency rate of the industrial process when it actually results from the combination of the two;

– the recycling rate is confused with the rate of incorporation of recycled materials into overall material production.

The utmost care must be taken when reading the literature to identify what lies behind a posted recycling rate.

The calculation of the different rates is based on the description of the flows (Figure 7.1).

The rates can be calculated for production waste streams only (symbolized by p), for end-of-life product waste only, which can be household or industrial (symbolized by XReol), or globally (symbolized by XRg).

The main rates calculated for waste recycling are as follows:

– the end-of-life waste collection rate (CReol):

$$CReol = f/e \quad [7.1]$$

– the efficiency rate of the end-of-life waste recycling process (EReol):

$$EReol = g/h \quad [7.2]$$

– the end-of-life waste recycling rate (RReol):

$$RReol = CReol * EReol \quad [7.3]$$

– the rate of incorporation of waste into production (IRg):

$$IRg = (i+j)/(a+i+j) \quad [7.4]$$

– the share of end-of-life waste in recycled waste:

$$Reol = g/(g+h) \quad [7.5]$$

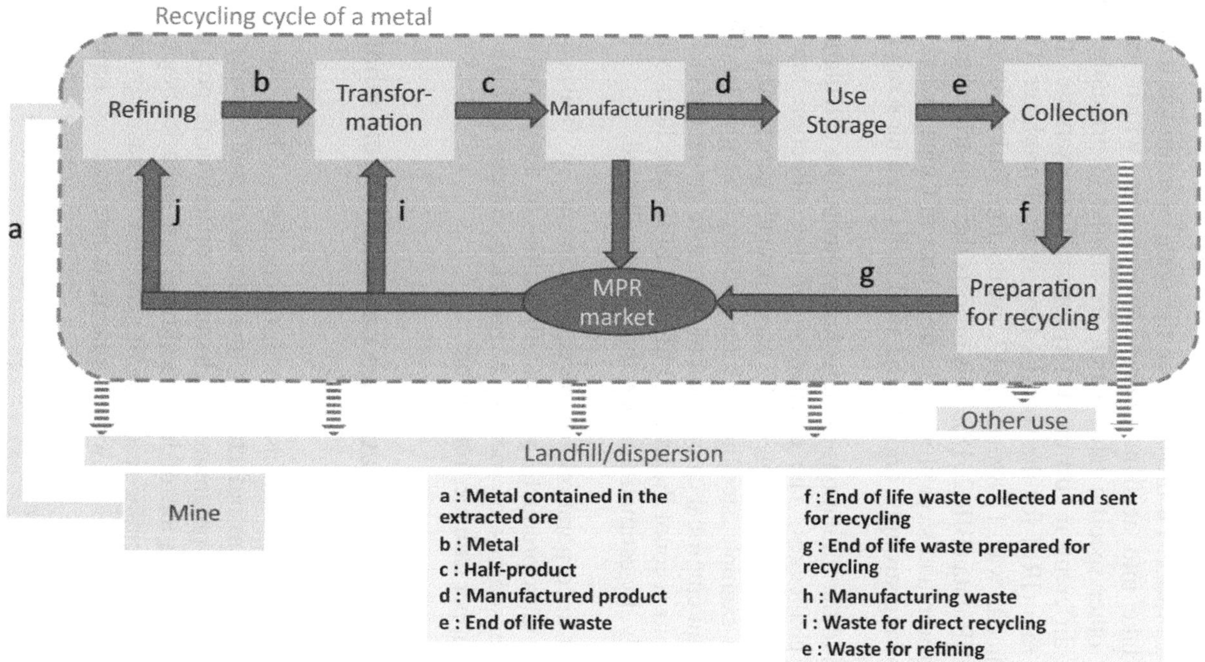

Figure 7.1. *Description of material flows in the metal cycle according to the UNEP (source: UNEP IRP (2011)). For a color version of this figure, see www.iste.co.uk/fizaine/mineral2.zip*

This diagram (Figure 7.1), describing a closed system, such as for the whole planet and for the usual flow of a metal, must be completed to take into account recyclables in a different sector, for example, copper in the manufacture of bronze. Furthermore, on the scale of a limited territory, it is necessary to take into account imports and exports, which leads to the calculation of other specific rates which are similar in principle. The rates can be very different in different territories. For example, the overall incorporation rate (IRg) for the world's largest steel producer, China, with 831 Mt produced, was 17.8% in 2017, while for Turkey it was 80.7%, according to BIR data (BIR 2018). This reflects the industrial structures of the countries, with iron and steel production from ores in China, whose economic demand is growing strongly, and electric steel mills feeding scrap to Turkey.

In addition to the limitations of statistical systems, the calculation of rates presents two challenges. The calculation of the collection rate generally comes up against the knowledge of the waste actually generated, part of which may be lost, for example, steel in reinforced concrete sent to landfills. More often than not, an empirical calculation of the waste generated can be carried out by taking into account the average lifespan of the products consumed or, more uncertainly, by estimating a part that is not collected.

The rate of incorporation, or recycled content, in finished products, particularly complex products, presents difficulties of determination because of the multiple sources of metals. A recycled copper cathode is totally identical to a copper cathode from an ore.

7.4. A limited deposit because we can only recycle what we have consumed

The waste stream of products and substances that we have consumed and discarded either during manufacture or after use, as a result of wear and tear or obsolescence, whether programed, fashionable or technical, progresses over time. Indeed, the evolution of humankind has led since its beginnings to a growing demand for metallic resources. It generates increasingly unacceptable environmental impacts, as shown by numerous international reports (UNEP 2013; OECD 2018), while fears of resource depletion are growing. Faced with this, the idea of a society based on the circular economy has taken shape. For this type of societal model, some people imagine that recycling would satisfy almost all needs: "it is enough to recycle everything to continue consuming". Unfortunately, this vision does not stand up to analysis, and a circular economy will not exist if the growth in demand remains significant. Indeed, we are faced with two constraints: on the one hand, a constantly growing demand for raw materials and, on the other hand, limited access to waste.

The growth in demand for base metals has been between 2% and 3% per year for more than a century. While this figure may seem low, it does lead to a doubling of annual quantities every 20 years or so. For specialty metals, the growth in demand is much higher, with, for example, 8% per year for cobalt over the last 10 years and probably even higher growth in the years to come with the development of electric mobility.

If the production waste that is immediately recycled does not generate additional demand for raw materials, the same does not apply to used products that result from the rejection of products stored in the technosphere for a variable period of time depending on the product.

In this context, studies have shown the theoretical limits of recycling supply by illustrating it for iron (Grosse 2010) or copper (Labbé 2016) (Figure 7.2). It is thus possible to calculate the maximum effect of recycling to satisfy demand without recourse to mining. The duration of use, and therefore of storage in the technosphere, is a determining factor in the availability of waste. It varies from less than a year for packaging to several decades for buildings and infrastructures.

Theoretically, with an average storage life (weighted average of the lifetimes of the different uses) of copper of 35 years, a growth rate in demand of 2.87% and an end-of-life recycling rate (EOL-RR) of 80%, recycling will not be able to satisfy more than 30% of demand. For copper, out of the approximately 700 Mt of copper mined since the beginning of humankind, 452 Mt were in 2017 considered still in use and 12.5 Mt generated waste in 2017 (ICSG 2018). Currently, with an end-of-life recycling rate of 48%, the coverage rate is 20%.

In addition, waste cannot be 100% recycled for practical, technical and economic reasons. While metals can be recycled ad infinitum, there is a loss in each cycle. For every 100 t of metal that is 80% recycled, after five cycles, just over 32 t will remain.

The end-of-life recycling rate varies greatly depending on the metal. From 84% for steel (Russo 2018), it is almost zero for many specialty metals used in new technologies (UNEP IRP 2013).

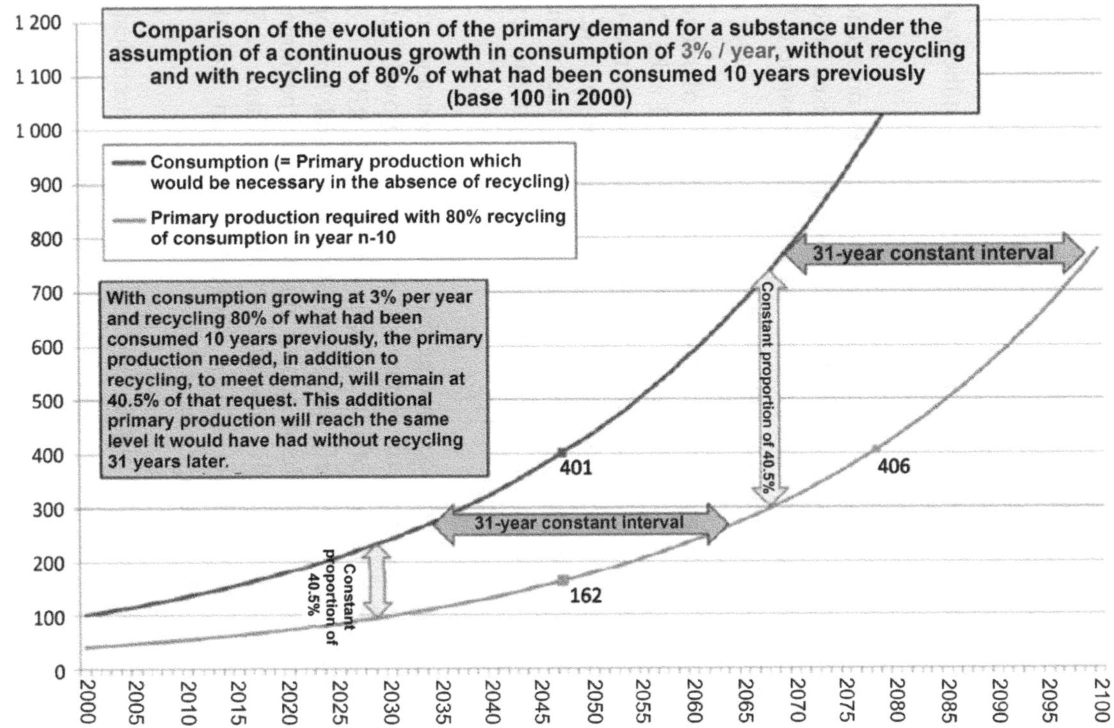

Figure 7.2. *Example of the relationship between material consumption and the contribution of recycling (source: Labbé (2016)). For a color version of this figure, see www.iste.co.uk/fizaine/mineral2.zip*

This, for specialty metals, means that their share of supply through recycling will remain low for a significant period of time until used products are replaced in large volumes. This is the case, for example, for rare earths, which will only be able to represent a significant volume in terms of recycling after 2030 (Rademaker *et al.* 2013) in the face of a demand that will have increased at the same time.

We can say, in a rather colorful way, that recycling will only chase a growing demand without ever satisfying it. This, of course, raises the question of this growth in demand and of our models of society in a finite world, as raised in the Club of Rome report as early as 1972 (Meadows *et al.* 1972, 2012).

7.5. Multiple factors influencing recycling and its effectiveness

The recycling chain is made up of a set of successive and parallel stages, all the more numerous as the waste is complex. If for unpolluted production waste and in significant quantities the circuit can be summed up as a transport phase from the waste generator to a direct loading at a semi-product producer, for waste from used products it is necessary, on the one hand, to collect this waste and, on the other hand, to prepare the materials through a series of steps in the recycling chain (Figure 7.3).

It is important to understand that each step induces an unavoidable loss of material to a greater or lesser extent in residues, which in most cases will not be reprocessed. Figure 7.3 shows an estimate of the current average efficiency rates of each step.

Recycling, and first and foremost collection performance, is influenced by the local political, cultural, social and economic contexts. The political framework, translated into legislation and regulations, can lead to strong recycling obligations involving the most efficient collection possible. This is notably the case in Europe and more particularly in France with the significant development of extended producer responsibility (EPR) schemes for end-of-life products (ADEME 2017) since the early 1990s. These regulations set ambitious collection and recycling targets for product categories. Japan, with its low mineral resources, its 3Rs policy with international reach (Japanese Ministry of the Environment 2005) and its legislation aimed at a "material-cycle friendly society", has made recycling an important part of its industrial supply. In addition, China enacted a circular economy law in 2006 that strengthens the place of recycling. Other developed countries, such as the United States, are less involved in these regulations.

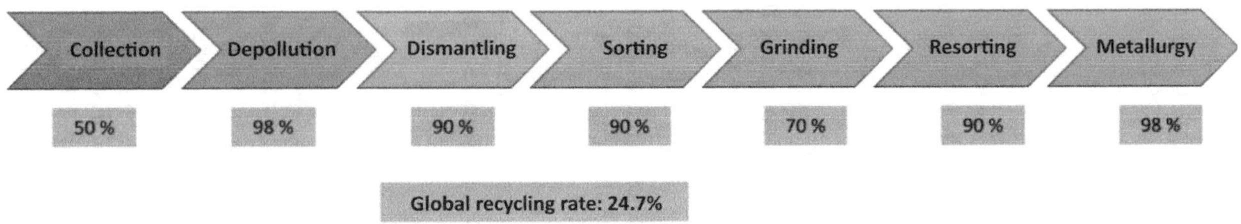

Figure 7.3. *Illustrative recycling chain stages and efficiency rates. For a color version of this figure, see www.iste.co.uk/fizaine/mineral2.zip*

However, the objectives of these regulations are all global and lead to the recycling of base metals without any specific focus on low-concentration metals, except for precious metals carried by their high value.

The recycling potential is also driven by the use of metals and substances. Some substances are dispersed by their use, such as copper sulfate in agriculture and titanium oxide, used as a white colorant in cosmetics, food or paints, which cannot be recycled. Conversely, manufacturing waste, both internal and new scrap, is heavily recycled. Certain manufacturing processes generate, by their very nature, significant proportions of waste. The manufacture of flat screens uses ITO (indium and tin oxide) with an efficiency of 15%, i.e. with 85% of the waste available for recycling. Similarly, the manufacture of forged titanium alloy parts (TA6V alloy for aeronautics in particular) generates 80% waste, which is largely recycled in a specific European-scale plant in France, EcoTitanium, in the Puy de Dôme region.

The potential value of waste is a determining factor in its recycling. The interest for recyclers is constrained between two competitive limits: on the one hand, the price of the material from ores and, on the other hand, the cost of landfilling the waste. For this second terminal, European regulations aimed at limiting or even prohibiting landfilling and the difficulty of opening new landfills tends to make it more expensive, thus encouraging recycling.

Depending on the country, the recycling chain can be notoriously different (UNEP IRP 2013). In developed countries, collection, which is regulated, must be profitable for operators of a certain size; pre-treatment is mechanized, with, if necessary, a depollution phase; metallurgical treatment is integrated with large smelting plants and emissions treatment. In developing countries, collection is informal but can lead to the collection of very diffuse waste in small quantities for which no clean-up will be provided; pre-treatment is often manual and metallurgy is artisanal, sometimes even hazardous where chemical treatments are carried out in backyards. This situation generates pre-judicial working conditions for workers, has a strong impact on the environment and competes with industrial facilities with high environmental quality. In addition, the efficiency of recycling is much lower.

Informal collection and the economic difficulties of recycling for metals may also have existed in developed countries not so long ago, including in France. A study carried out by ADEME on lead batteries (ADEME 2012) has shown that in 2000, 67% of the collection was informal but with a profit of 9.15 €/t while the organized collection without invoicing was at a loss of 3.05 €/t. The collection with invoiced services showed a profit of 62.51 €/t.

Various factors impact the development of pre-treatment and treatment facilities:

– the price of raw materials, a determining factor which, when markets are under pressure, not only accelerates recycling opportunities, but also, due to their variability, weakens the sustainability of investments. Price variability is a major obstacle to initiative and innovation;

– the knowledge, or rather the lack of knowledge, of the variable content of the waste at the entrance of the installation, required in order to adapt or optimize the process. This is an important issue requiring the development of high-performance analytical methodologies, if possible on a continuous basis;

– the evolution of products and therefore waste over time, which can lead to a lack of perspective for some facilities. Thus, the rapid replacement of compact fluorescent lamps by LEDs has not only limited in time the availability of the deposit but also the outlet for rare earths that can be recycled;

– the small quantity involved for some metals, a few hundred tons worldwide, which does not offer an interesting potential of value given the costs of treatment or the critical size of the necessary installations;

– if recycling generates employment, 15–30 jobs per 10 kt/year processed (ADEME 2014), it is therefore dependent on the cost of labor as well as the cost of energy. The OECD estimates that by 2060 the development of recycling will be hampered by labor costs (OECD 2018);

– the difficulty of raising investment funds that seek short-term profitability, whereas processes, excluding those for base metals, are often innovative and require a significant amount of industrial development time.

7.6. The technical constraints of metal recycling

If, contrary to organic carbon chains, the recycling of metals is theoretically infinite, because the atom is not affected by the processes, various factors lead to the limiting of the conditions of extraction and reuse of the material resulting from recycling.

Metals have different recycling characteristics depending on whether they are base metals, which have structural uses because they can be used for large parts and are present in amounts of several millions of tons, or specialty metals for which only between a few hundred thousand tons and a few hundred or even tens of tons are used worldwide. The latter are substances used either in very small quantities for their own metallurgical or electronic properties or as alloying elements in base metals.

7.6.1. *Preparation of materials*

Upstream of the metallurgical process, after possible decontamination of the used equipment, such as the removal of fluorinated gases, PCBs, etc., a crucial separation step is carried out to concentrate the components and direct them to the appropriate dies. This generally involves producing separate streams of glass, plastics, base metals and complex components for further processing, usually by other operators.

Manual dismantling can be carried out, but this involves a high labor cost although it has a high qualitative efficiency. It is generally essential to valorize large equipment, such as televisions and screens, and partially for vehicles. For equipment such as small electrical appliances and for carcasses resulting from manual dismantling, mechanical separation techniques using impact or crushing are most often used. The stakes and limits of this stage concern the selectivity of the separation limiting the mixing or degradation of the materials that can lead to a loss of material as dust (Table 7.2). The size of the resulting pieces must allow efficient recognition in the next step while having a good separation.

	Silver g/t	Gold g/t	Palladium g/t
Electronic cards, not shredded	669	135	50
Shredded electronic boards less than 8 mm in size	562	126	48
Shredded electronic boards less than 2.5 mm in size	481	48	18

Table 7.2. *Concentrations of precious metals in the stream to be recycled according to the level of electronic card shredding (source: Chancerel* et al. *(2009))*

This separation is followed by a sorting phase that can be based on different techniques, essentially an automatic one, but it does not exclude a manual finishing phase, as for the removal of electric motor windings from steel flows. The techniques use either visual aspects, such as shape, color and size; densimetric properties, determined through flotation, vibration or cycloning; magnetic or electromagnetic properties; or X-rays, infrared, plasma atomic emission or neutron gamma metrology. The efficiency of these techniques depends on the materials and their sizes, for which multi-centimeter fractions are often necessary. A technique will only be operational if it allows separation rates of several tons per hour with a separation rate higher than 90%. This good separation is necessary for the quality of the metals resulting from recycling. For example, the accumulation of copper due to the presence of electric motors, whose steel component makes the coil attracted to the steel, can lead to a loss of quality of the steel produced and implies for steelmakers that they must dilute batches rich in copper with poor batches.

At this stage the process is generally optimized to extract the base metals and can be carried out in successive stages, possibly by different operators. This is, for example, the case for some scrap metal shredding yards where the non-ferrous metals are separated in a mixture by other operators.

The separation is not perfect: there are release imperfections resulting from too much interweaving of the materials. Solving this problem would require a very fine release mesh which, on the one hand, is costly in energy, and, on the other hand, detrimental to certain stages of the treatment. There are also separation imperfections because automatic sorting techniques can lead to errors.

7.6.2. *Recycling of base metals*

Base metals are rarely used pure, except for copper in the electrical and plumbing industries with grades greater than 99.3% (Copper Alliance 1997) and 99.9% (Copper Initiative 2019), respectively. The compositions of alloys, whether ferrous or non-ferrous, include many additional metals that can be added to their composition in varying degrees of variety or quantity. Cast iron contains up to 4% carbon and low carbon steel is available in over 5,000 different alloys with many additional metals in the same alloy totaling less than 4%. In the same way, aluminum, zinc, titanium, etc., are in alloyed form, with sometimes very mixed compositions, such as bronze, a mixture of copper and tin, or brass, a mixture of copper and zinc.

This very large variety of alloys, possibly several tens of thousands, is a determining factor in the efficiency of metal recycling, with two main consequences. The first is the obligation to choose to sort the alloys in the recycling chain by their nature in order to recycle them separately and keep all their value. This is the case for stainless steel, with more than 10.5% chromium, and common steel, with less than 4% alloying metals, or for aluminum with aerospace alloys. If the choice is made not to sort the alloys, then a mixture with average characteristics is obtained, for use in low-stress applications, which is the case for steel, which leads to the production of long products, in particular concrete reinforcing bars, or aluminum for the production of automobile crankcases. The second consequence involves the loss, not physical but of use, of alloy metals, which is referred to as non-functional recycling. Indeed, it is impossible, for thermodynamic and thus cost reasons, to remove these alloy metals from the melting baths by the usual industrial processes. Recycling is then carried out for these metals mainly by melting them in a secondary melting furnace.

7.6.3. *Recycling of specialty metals*

With the exception of alloys – for example, beryllium in copper from electrical contacts for aeronautics, which is lost – specialty metals can be recycled, particularly from electrical and electronic scrap. Recycling here is more complex because it requires determining the different physico-chemical properties of the metals to separate and purify them. It is most often aimed at obtaining, *in fine*, very pure metals or metal compounds of 99%–99.999% purity. The processes frequently combine pyrometallurgy and hydrometallurgy in a succession of stages in installations requiring high investment but also, above all, know-how acquired through experience (Figure 7.4).

The treatment processes allow above all the recovery of most of the value produced by copper, gold and palladium (Table 7.3). The pyrometallurgy part separates the metals into three phases, the metal bath, the slag and the gases. Each of these reprocessed phases makes it possible, depending on the technological choices, to extract a certain number of metals with an efficiency generally higher than 95%.

However, depending on the content and physico-chemical behavior, it may be either impossible to extract certain metals or necessary to choose not to. This is currently the case for tantalum and rare earths. However, some residues from these processes can concentrate certain metals sufficiently for them to be sent to specialized stakeholders.

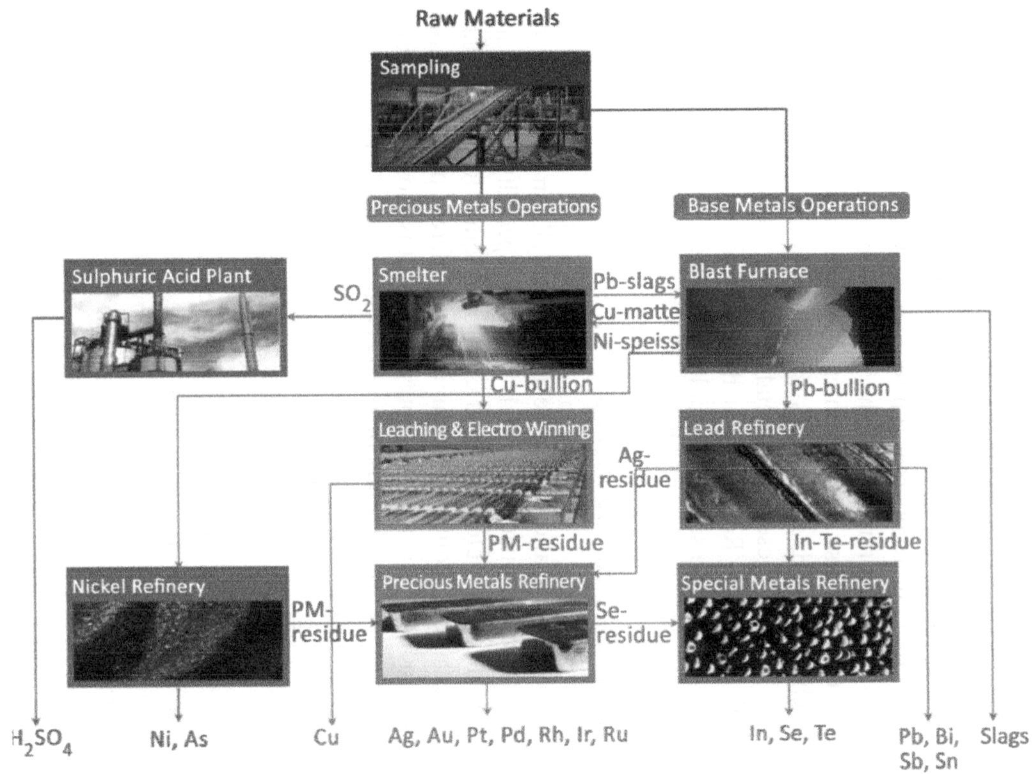

Figure 7.4. *Diagram of the electronic waste treatment process from UMICORE to Hoboken Belgium (source: Bouillon and Havard (2011)). For a color version of this figure, see www.iste.co.uk/fizaine/mineral2.zip*

	% of revenues
Gold	50.4
Copper	13.9
Palladium	9.5
Plastics	9.2
Silver	3.6
Aluminum	2.5
Pewter	2.0
Barium	1.8
Platinum	1.7
Cobalt	1.6

Table 7.3. *Distribution of income in the electronic waste deposit (source: Cucchiella* et al. *(2015))*

7.7. Environmental benefits of recycling

Recycling has various environmental benefits. The first relates to the limitation of mining extraction and the resulting impacts: damage to countries, water pollution and effects on biodiversity. The second is lower energy consumption, the possibility of using electricity with lower greenhouse gas emissions. Recycling steel saves 58% in greenhouse gas emissions, while for aluminum the saving is 93% compared to virgin aluminum (FEDEREC & ADEME 2017). The overall importance of the environmental benefits of recycling can be estimated by monetizing the externalities[3]. Such an exercise was conducted by the French Ministry of the Environment in 2007 for four materials, including aluminum (French Ministry of the Environment 2007). This work resulted in an economic differential in favor of recycling of 2,200 €/t for a price of 1,664 €/t at that time, close to the current price.

3 The impacts taken into account in this study were: air acidification, the greenhouse effect, the destruction of stratospheric ozone, degradation of water quality, human toxicity, toxicity to ecosystems, consumption of natural resources and various nuisances, including those suffered by humans and damage to structures.

However, it should not be forgotten that, like any industrial and especially metallurgical facility, the metal recycling industry requires special attention to limit potential local impacts. Given the presence of halogenated organic compounds that can emit dioxins and furans, gas or dust emissions that may contain heavy metals, extensive flue gas treatment is necessary. The same applies to liquid emissions from hydrometallurgical processes. The metal recycling industry is not exempt from damage to the environment around the installations. While this situation is well-controlled in most European installations, this is not the case in all countries.

7.8. Conclusion

From a subsistence or scarcity economy, recycling has become a political opportunity for governments: on the one hand, to respond to environmental issues and, on the other hand, to respond to economic issues, such as the development of activity and employment and also the need to ensure independence from mining production that is not present on the territory. The evolution of recycling is underpinned by the adaptability of this sector to processing used products, which are evolving more and more quickly, becoming more complex and, moreover, for technical products, reducing the recycling potential through miniaturization. The economy of materials, the use of more nested and complex materials, including for environmental reasons, may come up against recyclability. Entire sectors of the recycling industry may find themselves concerned by profound technological changes due to the lack of outlets for substances that are no longer used. In the recent past, the transition from cathode-ray screens to flat screens has rapidly compromised the recycling of lead and barium glass. Aluminum recyclers are currently wondering about their future in the face of a more or less rapid replacement of internal combustion vehicles by electric or hydrogen powered vehicles. The main outlet for recycled aluminum is in the engine blocks of internal combustion vehicles, which are less demanding than other outlets.

The strengthening of environmental protection policies around the world will intensify interest in recycling. However, in the absence of a significant increase in mining raw material prices due to possible shortages, recycling will only reach its full potential through policies that set quantified recycling obligations for each metal or family of metals. The use of economic instruments such as taxation, extended producer responsibility or recycling certificates can also be considered.

The development of waste collection is a prerequisite for improving recycling rates. Technical improvements can improve dismantling and sorting. Efficient and resilient processes with small units need to be developed for technological metals.

In the short and medium terms, recycling, particularly of metals, given the lifespan of the products concerned and the growth in demand, can only represent a limited but essential response to demand, particularly for specialty metals. In the longer term, with the saturation of developing economies (Bleischwitz and Nechifor 2016), the stagnation or even decrease in demand would lead to an increase in the role of recycling, which could become dominant while retaining a share of mining to compensate for the inevitable loss during the recycling process. At this point, metallurgy would probably have to turn to so-called molecular recycling, allowing the decomposition of alloys (Niesing 2013).

Recycling is part of a sustainable waste policy framed by the notion of hierarchy, which puts waste limitation at the highest level above reuse and repair and above recycling. Recycling thus appears to be the worst of the good solutions. The more recent notion of the circular economy has taken on a dimension of efficient resource management which, in order to make sense, must be part of strong limitations on resource consumption.

7.9. References

ADEME (2012). Audit économique de la collecte et du recyclage des batteries au plomb. ADEME, Paris.

ADEME (2014). Positionnement et compétitivité des industries de recyclage en France. ADEME, Paris.

ADEME (2017). Les filières à responsabilité élargie du producteur : mémo 2017. ADEME, Paris.

Amick, D. (2014). Reflection on the origins of recycling: A paleolithic perspective. *Lithic Technology*, 39(1), 64–69.

Barles, S. (2005). *L'Invention des déchets urbains*. Champ Vallon, Ceyzérieu.

Béguin, M. (2013). L'Histoire des ordures : de la préhistoire à la fin du dix-neuvième siècle. *Vertigo*, 13(3).

Bleischwitz, R. and Nechifor, V. (2016). Saturation and growth over time: When demand for mineral peaks [Online]. Available at: https://www.centre-cournot.org/img/pdf/prisme_fr/Prisme%20N°34%20Novembre%202016%20(english).pdf [Accessed May 2019].

Bouillon, C. and Havard, M. (2011). Gestion durable des matières premières minérales. Information report, Assemblée nationale, Paris.

Center for Archaelogy (2001). Archaeometallurgy Guidelines. English Heritage, Swindon.

Chancerel, P., Meskers, C.E.M., Hagelüken, C., Rotter, V.S. (2009). Assessment of precious metal flows during preprocessing of waste electrical and electronic equipment. *Journal of Industrial Ecology*, 791–810.

Copper Alliance (1997). Innovations, the metallugy of copper wire [Online]. Available at: https://www.copper.org/publications/newsletters/innovations/1997/12/wiremetallurgy.html [Accessed 5 June 2019].

Cucchiella, F., D'Adamo, I., Koh, L.S.C., Rosa, P. (2015). Recyclings of WEEEs: An economic assessment of present and future e-waste streams. *Renewable and Sustainable Energy Reviews*, 263–272.

EUROSTAT (2019). Contribution of recycled materials to raw materials demand [Online]. Available at: https://ec.europa.eu/eurostat/data/database?node_code=cei_srm010 [Accessed 8 June 2019].

FEDEREC and ADEME (2017). Évaluation environnementale du recyclage en France selon la méthodologie de l'analyse de cycle de vie [Online]. Available at: https://presse.ademe.fr/wp-content/uploads/2017/05/FEDEREC_ACV-du-Recyclage-en-France-VF.pdf [Accessed May 2019].

French Ministry of the Environment (2007). Monétarisation des impacts environnementaux du recyclage : méthodologie et applications [Online]. Available at: http://temis.documentation.developpement-durable.gouv.fr/docs/Temis/0062/Temis-0062814/18016.pdf [Accessed May 2019].

Geldron, A. (2016). Métaux stratégiques : la mine urbaine française. *Responsabilité et environnement*, April, 67–73.

Graedel, T.E., Allwood, J., Birat, J.-P., Buchert, M., Hagelüken, C., Reck, B.K., Sibley, S.F., Sonnemann, G. (2011). What do we know about metal recycling rates. *Journal of Industrial Ecology*, 355–366.

Grosse, F. (2010). Le Découplage croissance/matières premières. *Futuribles*, July–August, 100–124.

ICSG (2018). The World Copper Factbook 2018. International Copper Study Group, Lisbon, Portugal, 66.

Initiative Cuivre (2019). Mise en œuvre – normes [Online]. Available at: https://tube-cuivre.fr/mise-en-oeuvre/normes [Accessed 5 June 2019].

Japanese Ministry of the Environment (2005). *Japan's Experience in Promotion of the 3Rs*. Japanese Ministry of the Environment, Tokyo.

Kuijper, M. (2019). Circular economy: Ancient populations pioneered the idea of recycling waste. *The Conversation* [Online]. Available at: https://theconversation.com/circular-economy-ancient-populations-pioneered-the-idea-of-recycling-waste-107332 [Accessed 21 May 2019].

Labbé, J. (2016). Les limites physiques de la contribution du recyclage à l'approvisionnement en métaux. *Annales des Mines – Responsabilité et environnement*, 45–56.

Meadows, D., Meadows, D., Randers, J., Behrens III, W.W. (1972). *The Limits to Growth*. Universe Books, New York.

Meadows, D., Meadows, D., Randers, J. (2012). *Les limites à la croissance (dans un monde fini)*. Rue de l'échiquier, Paris.

Niesing, B. (2013). Recycling 2.0. *UmweltMagazin*, April, 12–16.

OECD (2001). *Responsabilité élargie des producteurs : manuel à l'attention des pouvoirs publics*. OECD, Paris.

OECD (2017). *Responsabilité élargie du producteur : mise à jour des lignes directrices pour une gestion efficace des déchets*. OECD, Paris.

OECD (2018). *Global Material Resources Outlook to 2060: Economic Drivers and Environmental Consequences*. OECD, Paris.

Rademaker, J., Kleijn, R., Yang, Y. (2013). Recycling as a strategy against rare earth element criticality: A systemic evaluation of the potential yield of NdFeB magnet recycling. *Environmental Science & Technology*, 10129–10136.

Russo, P. (2018). Le recyclage de l'acier un atout pour la société. *Atelier thématique annuel EcoSD*.

Shimelmitz, R. (2015). The recycling of flint throughout the lower and middle paleolithic sequence of Tabun Cave, Israel. *Quaternary International*, 361, 34–45.

UNEP (2013). Environmental risks and challenges of anthropogenic metals flows and cycles [Online]. Available at: https://www.resourcepanel.org/reports/environmental-risks-and-challenges-anthropogenic-metals-flows-and-cycles [Accessed May 2019].

UNEP IRP (2011). Recycling rates of metals: A status report [Online]. Available at: https://www.resourcepanel.org/reports/recycling-rates-metals [Accessed May 2019].

UNEP IRP (2013). Metal recycling: Opportunities, limits, infrastructures [Online]. Available at: https://wedocs.unep.org/bitstream/handle/20.500.11822/8423/Metal_Recycling-FReport.pdf?sequence=1&isAllowed=y [Accessed May 2019].

8

Low-tech: A Path Toward the Necessary Metallic Sobriety?

Philippe BIHOUIX
AREP, Paris, France

8.1. Cornucopians versus doomsdayers

The last decade has seen the resurgence of an old debate on the issue of resources, particularly "non-renewable" resources such as fossil fuels and metals. This debate was lively from the late 1940s to the early 1980s: through scientific publications or best-selling books, the neo-Malthusians, described by their detractors as "doomsdayers", opposed the "cornucopians"[1].

The former, often coming from the natural sciences (Osborn 1948; Ehrlich 1968; Vogt 1948), loudly proclaimed that the combined effect of population growth and the development of consumption was bringing humanity to the brink, sooner or later. The latter, who were more from the social sciences, economists, futurists and prospectivists (Toffler 1970; Kahn and Wiener 1967; Nordhaus 1973; Nordhaus 1974; Simon 1981[2]), explained on the contrary that technological progress, increasingly rapid, would sweep away all the threats of penury (Bihouix 2019). Exchanges gained intensity with the

1 From the Latin *cornu copiae*, the horn of plenty.

2 Nobel Prize in Economics 2018 (Bank of Sweden Prize in Economics in memory of Alfred Nobel) with Paul Romer.

Mineral Resources Economics 2,
coordinated by Florian FIZAINE and Xavier GALIÈGUE. © ISTE Ltd 2021.

publication in 1972 of the report to the Club of Rome on the impact of economic and demographic growth by the Massachusetts Institute of Technology (MIT) team (Meadows *et al.* 1972).

The debate, however, faded in the 1980s. The risks of shortages seemed to be diminishing, as evidenced by the fact that the price of raw materials was permanently on the decline: the price of oil halved between 1984 and 1986 and remained low until 2003, when it began to make a spectacular recovery; the price of metals collapsed with the fall of the USSR and the marketing of strategic stocks from the two blocs, which became useless with the partial disarmament and then the end of the Cold War.

In the mid-2000s, concerns resurfaced. The notion of peak oil, theorized by the geophysicist Marion King Hubbert in the late 1950s, was revived by an association of former petroleum geologists[3]; initially rather confidential and informal, this time it was used to approach the next date of global peak oil. Futhermore, the accelerated development of emerging countries, especially China, put the markets for raw materials under pressure. Several of them (aluminum, copper, lead, zinc, silver, etc.) reached their record prices between 2007 and 2011, despite a temporary sharp decline following the financial crisis of 2008–2009.

Nowadays, the debate on the sustainability of our societies is raging again, and demographic issues are also beginning to resurface. Doomsdayers and scientific "whistleblowers" such as Edward Wilson for biodiversity, James Hansen for climate change and Jared Diamond – author, among others, of the remarkable book *Collapse* (Diamond 2005), which deals with the conditions and circumstances of the collapses of past civilizations – try to convince us that things are going to get worse and worse, perhaps even more abruptly. Researchers are not alone in sounding the alarm. Several official bodies, usually rather cautious and muffled in their communications, such as the United Nations Environment Programme, the World Bank and the Organisation for Economic Co-operation and Development, are concerned in recent reports about the sharply increasing drain on resources – global material extraction has tripled since 1970, and could rise from 70 billion tons in 2010 to 180 billion in 2050 – and the increased need for scarcer raw materials generated by the fight against climate change (World Bank 2017; OECD 2018; UNEP 2019).

3 The Association for the Study of Peak Oil (ASPO).

There are also many cornucopians, including prospectivists, like Jeremy Rifkin, advocate of the Third Industrial Revolution (Rifkin 2011), and also emblematic figures, not at all frightened by amazing technological prophecies: the multi-entrepreneur Elon Musk (PayPal, Tesla, Space X, Neuralink, The Boring Company, etc.) has ambitions to revolutionize renewable energies and electric transportation, and is already proposing to terraform Mars; billionaire Jeff Bezos, founder of Amazon, sees no limits to the galactic expansion of the human species, which could then, without problem, reach the figure of 1,000 billion individuals; Google is multiplying research on artificial intelligence and the means to extend lifespan, and funds the Singularity University of Ray Kurzweil[4], recruited to its management team in 2012.

Between the risk of collapse and a radiant future, who should we believe? At the heart of the debate, then as now, is the question of technological progress – and it must be acknowledged that it has been impressive over the last few decades. According to optimists, it should make it possible to move from environmentally destructive growth (the current degradation is difficult to contest) to a "green" growth (more) respectful of ecological balances, and even restorative, with the possibility of cleaning up the oceans and the soil, restoring ecosystems, or even, in the most forward-looking visions, taking charge of global climate regulation through different geo-engineering techniques, even reviving certain extinct species.

From this point of view, the question of non-renewable resources remains – since they are drawn from a stock that is by nature limited, however large it may be – but it is generally dismissed by the cornucopians: fossil energies will be replaced by renewable energies (and/or "sustainable" nuclear energy, such as fast breeder reactors or perhaps fusion[5]); metals will either be saved at the source (reduction of needs through eco-design, new processes that are more efficient and therefore less consuming, replacement by certain renewable materials: biomaterials, biocatalysis, etc.), or reused or recycled in a circular economy approach.

4 See the article "Technological Singularity" on Wikipedia.

5 The reserves of uranium 235, the fuel for the second- and third-generation power plants, do not go beyond the century, at the present rate of use. Breeder reactors with uranium 238 (fourth generation power plants), or fusion (based on deuterium/lithium, for example) would offer prospects over a longer period (perhaps a few thousand years, depending on the quantity of reactors deployed).

In any case, the need to enter into a more circular logic, to "dematerialize" the economic system, to decouple economic growth from the consumption of resources, is recognized by everyone. It is true that without this decoupling, long-term projections would quickly become absurd when confronted with the necessarily limited physical reality. A growth rate of 2% per year (lower than the growth rates of energy or resource consumption in recent decades[6]) means doubling every 37 years, multiplying by 7 in a century, and by 390 million in a millennium. In other words, it would take about 400 years at this rate to multiply our energy consumption by 10,000 and consume all the energy from the Sun arriving on the Earth – from an instantaneous power of about 16 TW[7] to 170,000 TW. It would then take only 1,164 years to multiply this figure by 10 billion and consume all the energy of the Sun, which would make our civilization a "Type II" on the scale of the Soviet astronomer Nikolai Kardachev[8].

With the same type of multiplicative factors, resources would quickly be limited by the same physical constraints: it would take eight to nine centuries to consume the entire iron content of the asteroid belt (Elvis and Milligan 2019) and five centuries to extract the entire copper content of the Earth's crust to a depth of 1 kilometer[9]. Without decoupling, the infinite growth hoped for by economists would turn into farce or science fiction. It will therefore be necessary to move from a physical economy to a "knowledge" economy, where the "combinatory" explosion of ideas could continue (almost) indefinitely: "at no conceivable horizon – I would say until about five billion years from now, when the Sun explodes – will we run out of new discoveries", prophesied the 2018 Nobel Prize winner in economics[10] Paul Romer (Bailey 2001).

8.2. The circular economy, mission impossible?

Considerations on the possibility of dematerializing the economy are not new. At the end of the 1930s, the architect Buckminster Fuller spoke of the concept of ephemeralization (Fuller 1938), the fact that fewer and fewer raw

6 See below.

7 1 TW (terawatt) = 1,000 gigawatts or 1,000 billion watts.

8 See the article "The Kardachev scale" on Wikipedia.

9 In the order of 20,000 billion tons (average concentration of 50 parts per million, 2.6 tons per m^3, on 150 million km^2 of land), compared to a current production of 20 million tons.

10 See note 2.

materials were being consumed in order to render equivalent or better services; radio waves thus replaced the copper cables of the telegraph and, with the geodesic dome structure of his invention, the quantities used in buildings could be considerably reduced.

The argument is still being used today: optical fibers, which are lighter and made from more abundant materials[11], carry much more information than copper cables; the miniaturization of electronic objects, or their multiple functionalities (the smartphone that also has a camera, GPS and flashlight, etc.), reduces resource consumption, etc. More generally, everyone agrees that, as economies develop, the weight of industry, which consumes materials, is diminishing in relation to services, intangibles and knowledge.

8.2.1. *Invisible dematerialization*

However, behind the concept, the figures are less eloquent. If we do indeed see a relative decoupling, a disconnection between gross domestic product (GDP) and energy consumption[12] or CO_2 emissions, it is an absolute decoupling that we would have to achieve, with a GDP that continues to grow while emissions fall. The same result should be obtained for metal resources, where to date no decoupling, even relative, is visible at the global level. Between 2001 and 2016, mining and metallurgical productions have grown each year in a comparable manner and often more than the GDP for almost all metals: iron (4.6%), aluminum (5.9%), copper (2.3%), lead (2.9%), nickel (3.5%), cobalt (8.4%), etc. Over the last five years, the phenomenon has become more pronounced: 3.6% for copper and lead, more than 7% for aluminum and nickel (US Geological Survey 2019) (Figure 8.1).

Certainly, we need to look beyond the global figures: in Europe, which is technically more "advanced", CO_2 emissions are stagnating or falling slightly; the industrialization, urbanization and motorization phase in emerging countries, led by China and India, may explain the growth in global demand for metals and the additional extraction needs, while in more developed countries, recycling is gradually increasing and should contribute to limiting mining demand in the future.

11 Although they also contain rarer products, such as erbium (a lanthanide), to increase the refractive index.

12 Between 2001 and 2016, world GDP grew by an average of 2.8% per year, and CO_2 by "only" 2.1% per year; over the most recent third of the period (2011–2016), GDP was still at 2.7%–2.8%, but CO_2 at 0.6%.

However, we have to take into account the effect of the balance of foreign trade: if Europe emits less CO_2, for example, it is because it is imported in the form of finished and semi-finished products. The French Commission on Sustainable Development (CGDD) noted an increase in the carbon footprint of 5% per French person from 1990 to 2007, while the average level of emissions per person within the territory decreased by 15% (CGDD 2012). The same observation applies to raw materials: the decoupling between growth and the mobilization of materials is "partly the result of a transfer of extractive and industrial activities to emerging and/or developing countries" (CGDD 2018).

Our societies have never been so material, they have even become "hyper-industrial" (Veltz 2017), consuming more resources and energy than ever before and emitting more greenhouse gases than ever before.

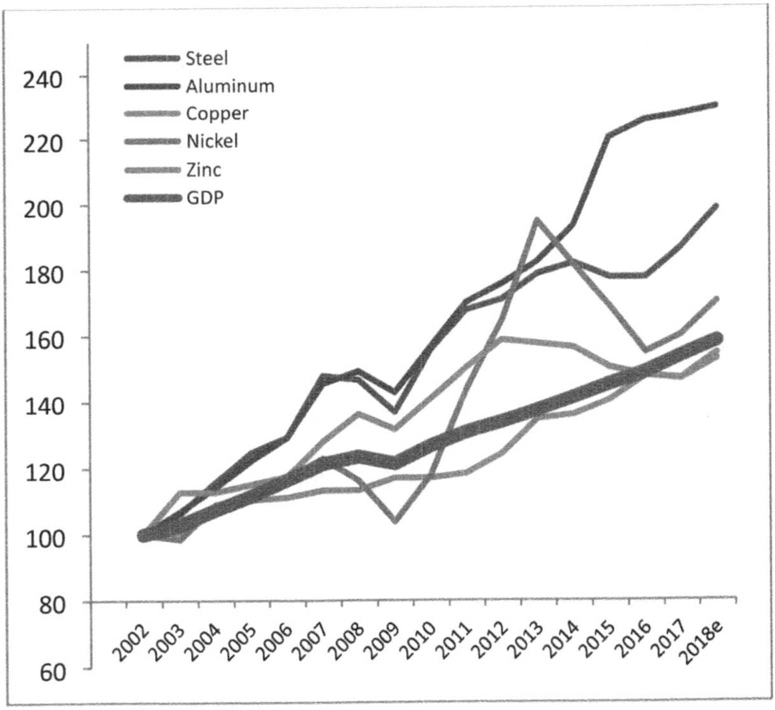

Figure 8.1. *Comparative change in world gross domestic product (GDP), world steel production and mining production of selected major industrial metals over the last 15 years (base 100 in 2002) (sources: World Bank data and US Geological Survey mineral commodity summaries, 2003–2019; production for steel and mining for other metals). For a color version of this figure, see www.iste.co.uk/fizaine/mineral2.zip*

To feed the "boiler" of growth, we have to dig. Always more tons, always deeper, with more and more powerful means, in more and more remote places, etc. In 25 years, we have almost tripled the tonnage extracted each year of iron, aluminum and nickel and more than doubled that of copper, zinc, lead, etc. Some metals used in new technologies (cobalt, indium, lithium, etc.) have experienced even faster growth, driven by digital applications.

8.2.2. *The systemic issue between energy and resources*

In addition to the "extractivism" of fossil energy resources, there is also a metallic "extractivism" that has nothing to envy – given the growth rates of the last decades, the cumulative extraction of each of the metals[13] will be more important over the next 20–30 years than the quantity extracted from the end of the Neolithic period to the present day.

The latter could be exacerbated by a massive program of equipment in renewable energies and "decarbonized" individual mobility devices. Indeed, while renewable energies are theoretically inexhaustible and available in large quantities – at least as long as the growth in demand is not prolonged over several centuries[14] – devices, notably made up of metals (and often fossil fuel resources), are nevertheless needed to capture, convert, store and use them (World Bank 2017).

Less concentrated and more intermittent, they produce less electricity per unit of metal (copper, steel, etc.) mobilized than fossil fuels. Some technologies use rarer metals, such as neodymium/dysprosium for high-power wind turbines, indium, selenium or tellurium for some high-efficiency PV panels[15], and other various metals for inverters, batteries, power electronics, etc.

Of course, technological advances can be made to reduce the amount of material mobilized or the dependence on the rarest chemical elements. However, there is no Moore's law[16] in energy: electrons, with mass, are more difficult to transport than photons, and energy is not stored as easily as bytes;

13 With the possible exception of mercury.

14 See above.

15 CIGS and Cd-Te technologies.

16 Empirical law, stated by Gordon E. Moore, one of the founders of Intel, postulate the doubling of the density of transistors every two years or so, making it possible to build ever more powerful and less expensive machines.

while efficiency can be improved, it is still necessary to use a certain amount of material capable of storing or exchanging electrons.

While there are significant mineral resources – exploitable reserves are increasing with technical innovations and lower production costs, or higher selling prices – their "quality" is declining, with the richest, most concentrated and simplest to extract resources being the first to be exploited. The new mines often have lower ore grades than the old ones, or are less accessible, deeper, and therefore more difficult to exploit. To extract the same amount of metal requires a greater amount of energy and technology – not to mention the environmental impact that could be multiplied, as in the case of possible underwater mines.

The availabilities of energy and metals are therefore strongly linked. If the problem was "only" that of fossil fuels and the climate, it would be possible to set up, as, for example, Jeremy Rifkin recommends, an "Energy Internet" based on solar panels and wind turbines, all regulated by smart grids, "intelligent" networks that balance intermittent supply and variable demand. If only metals were scarce, while access to concentrated and abundant energy was assured, it would be possible to exploit the Earth's crust at ever lower concentrations. However, isolating one of these problems is impossible, since they appear to be highly interdependent: if energy is used to extract and refine metals, the latter are also needed to produce energy more "cleanly".

8.2.3. *The constraints of recycling*

Will the demand for mining production eventually decrease and reach a more "sustainable" stage – before it is finally constrained by supply – with peak metals following or accompanying a peak in fossil fuel production? Metals, once used, do not disappear and can be recycled without losing their physical and chemical properties. However, such a circular economy would require a radical change in the current modes of design, production and consumption in order to really work.

In order to be recycled, the resource must first be physically recovered, which is not possible for dispersive uses of metals, incorporated as chemical products or additives in glass, plastics, inks, paints or cosmetics, among others: 5%–20% of many metals, such as chromium, cobalt, manganese, tin, zinc, molybdenum and lead, are used in this way. The next step is to recycle properly, without loss of functionality or quality (downcycling). This task is made difficult by the diversity and complexity of the materials used (composites, nanomaterials and alloys) and by the minute quantities of rare metals involved in ever more miniaturized and integrated components.

This phenomenon is particularly present in new technologies. For example, a smartphone contains 30–40 different metals, ranging from a few grams to a few tenths of a milligram or even less. The best recycling processes (such as in Hoboken near Antwerp or in Boliden in Sweden) for copper and lead/zinc can, through a complex sequence of hydrometallurgical and pyrometallurgical operations, recover only 10–15 of them[17]. Many "emblematic" high-tech metals, which are very costly to produce from an environmental point of view, are incorporated in products with a short lifespan, such as rare earths, tantalum, gallium or germanium, and are recycled at less than 1% worldwide (UNEP 2013).

8.3. Toward a metallic frugality

The more technologically complex the products and services around us and the manufacturing plants are, the further we move away from the circular economy and accelerate the rate at which non-renewable resources are drawn from the stock. Of course, we can continue to draw from the stock at an incredible rate, but we expose ourselves, on the one hand, to a medium- or long-term risk of "depletion" of resources (by the degradation of the metal content of the minerals or by the energy requirement generated by their exploitation, making their exploitation unprofitable from an economic and/or energy point of view), and on the other hand, in the shorter term, to increased environmental consequences (energy consumption, opening of new exploitation sites, etc.).

The other option would be to move as quickly as possible toward "metal frugality", a drastic reduction in the quantity extracted each year. A "low-tech" transition would not negate the possible interest of certain technological advances (it is not a question of going back to the candle), but would profoundly rethink innovation, focusing above all on sobriety, saving resources and recycling. This would involve questioning our needs whenever possible, pushing the logic of ecoconception much further than today, basing

17 Mainly the most precious: gold, silver, platinoids (platinum, palladium, rhodium and ruthenium), indium, selenium, tellurium and copper.

it, whenever possible, on sober, sustainable, agile and resilient technologies, and also addressing the organizational, social, societal, commercial, cultural or systemic components of innovation (La Fabrique écologique 2019; Bihouix 2020).

Given recent past growth rates, a 50%–65% decline would only bring us back to the extraction rate of the 1990s, which were not exactly "green" years. This gives an idea of the magnitude of the task ahead of us, when the population has grown in the meantime. Not to mention that exploiting "moderately" is not technically self-evident: with increasing depth, less accessibility or a decline in the metal content of ores, the mines of the future will require high capital expenditures and will generate ever-increasing operational costs, requiring massive and rapid exploitation that is not compatible with a decline in the volumes to be extracted in the long term.

8.3.1. *Sobriety above all*

How can the level of extraction be brought down to zero or almost to zero, corresponding to the logic of a truly circular economy, and thus be made respectful of its environment (reducing the impact) and of future generations (leaving a certain amount of stock in place for future needs)? This would probably start by avoiding the current acceleration, in particular that of an exacerbated and ubiquitous digital environment, associating connected objects, robots, drones, data centers for artificial intelligence software and storage of big data, and that of an irrational growth of renewable energies or electric and/or autonomous cars, without thinking about the indispensable sobriety that should be first implemented.

It would be a question, first of all, of reducing needs at the source, of working on the reduction of demand and not only on the replacement of supply, or, in other words, on sobriety before efficiency, two notions that are close, with an unclear boundary, and that we sometimes tend to confuse. Schematically, insulating a building to heat it less, or organizing carpooling, is about efficiency: the need is not questioned (here, living in a certain thermal comfort, or traveling a certain number of kilometers), but a more efficient offer makes it possible to reduce the "environmental bill". Sobriety would, in both cases, consist of questioning the need, by renouncing part of the thermal comfort or avoiding making the journey.

However, historically, efficiency gains that reduce unit consumption (of energy or resources) have, most of the time, been reduced or canceled out by the rebound effect - efficiency generally also translates into economic gain, then usage increases and therefore overall environmental impact increases. For example, the gains in kerosene consumption of the most recent turbojet engines (with rhenium alloy) have enabled the growth of low-cost aviation, the opening of new high-speed train lines does not empty planes but rather causes new journeys, and the architecture of data centers has reduced their energy bills, lowering the cost (per byte) of data storage, leading to a doubling of data produced, stored and exchanged every 18–24 months.

One can imagine a whole range of possible actions, more or less complicated, more or less acceptable, more or less quick to implement. Some of them could logically be agreed upon or almost agreed upon, provided that the arguments are well-explained: eliminating most disposable objects (except for certain medical uses), advertising materials, bottled water, etc. Others will be a little more difficult to get across because they require a change, a commitment, some "effort" and/or legal and regulatory adaptations: returning to bulk sales or charging for packaging, reusing objects, composting organic waste, limiting vehicle speed, etc. Finally, others promise some hot debates, and call for real systemic, societal, sociotechnical and cultural changes, such as reducing food waste, meat consumption or the "need" for hyper-mobility.

In the transportation sector, for example, there should be a gradual exit from the car civilization, which cannot be "sustainable" either from an energy or a resource point of view. The bicycle, with its thousand component parts, is a much more durable, sober, "autonomous" and easy to repair object compaed to a car full of electronics and high-tech metals which must meet anti-pollution standards. However, our urban planning, centered on the automobile for 70 years, does not allow everyone to ride a bicycle tomorrow for the majority of our journeys. This does not prevent us from preparing for this change by developing innovations in the field of two-wheelers and continuing our efforts in public transportation, while gradually lowering the weight and limiting the power and maximum speed of cars.

In the building sector, the need for new constructions could be reduced by intensifying the use of existing buildings to concentrate technical and financial resources on rehabilitation, renovation and embellishment. Schools, for example, are notoriously badly used outside of school hours, in the evenings, at weekends and during school vacations. Associations, clubs, events, etc., could be hosted there. Some alternatives emerge, such as architects' programs that mix functions, like a business platform during the day, a hotel at night, or shared housing with the pooling of certain rooms and equipment (laundry, games room and guest room). Numerous studies could be carried out to intensify the use of already built-up areas, rather than having to choose between urban sprawl and densification, each of which has its drawbacks. Then, of course, it is a matter of designing buildings that are as sober as possible, both in terms of construction (embodied energy, materials, etc.) and use (bioclimatic architecture, winter/summer thermal comfort, adaptability of needs, etc.). It is much simpler, much faster and much less costly in terms of physical and financial resources to insulate bodies than buildings – and there are now materials and techniques that would make it possible to do so in a fairly "comfortable" way.

8.3.2. *"Advanced" eco-design*

Second, our objects should be thoroughly rethought in order to considerably increase their lifespan and optimize resource recycling rates. Most disposable products or dispersive uses should be banned if they are not entirely based on renewable and non-polluting resources – which would in any case require a drastic reduction in the volumes used[18].

Sustainable products should be designed to be simple, robust and user-friendly, repairable and reusable, standardized, modular, based on simple materials and easy to dismantle and should only sparingly use non-renewable, therefore scarce and irreplaceable, resources. It would also be necessary to limit their electronic content, even to accept fewer demanding "specifications", lower performance, intermittency, loss of output and less "convenience" or increased maintenance requirements, but still retain the essential functionality.

Eco-design already integrates these concepts, but it should be pushed much further than it is today, as it faces three pitfalls. The first is methodological, because multi-criteria analysis (energy, materials, waste, etc.) is not always easy to handle. Is it better, for the planet, to consume and "lose" 5 grams more copper to emit 1 kilogram less CO_2, or to agree to emit a little more CO_2 to save scarce resources? The second is economic, because eco-design generally has a cost, and in the competitive game, the extra cost must be passed on to the customer, who must therefore be sensitive to the

18 Unless there are justified exceptions, such as in the medical field for example.

ecological argument, otherwise the company will lose market share. The third is philosophical, because the notion of "utility" does not appear in the analysis. Of course, this is not easy to grasp, but intuitively, it is clear that the consumption of scarce resources, for example, is more "admissible" or more "acceptable" for certain uses than for others. Even when eco-designed, the connected cat food dispenser, with a facial recognition device to avoid feeding all the cats in the neighborhood, still seems more "superfluous" (the superfluous admittedly being a "very necessary thing", as Voltaire said in the poem *Le Mondain*[19]) than a medical device – but what company starts by asking itself the question of the social utility of its products or services?

8.3.3. *Moderate mechanization*

Finally, we need to reflect on our production methods. Should we continue the race for productivity and the scale effect in giga-factories, or would it be better to develop workshops and companies on a human scale, at least for certain types of production? Should we not review the place of the human work, the sometimes unjustified degree of mechanization and robotization, the way we arbitrate today between manpower and consumption of resources and energy, at a moment in which we are talking about an ever-increasing number of jobs that could be replaced in the relatively near future by robots or artificial intelligence software?

There is also the question of production territories. After decades of globalization facilitated by sufficiently low transport costs, is the current system, based on China and Southeast Asia as "factories of the world" and complex, globalized flows, really resilient to the risks of future disruptions?

It is obviously not a question of "demechanizing" everything up to Gandhi's spinning wheel and animal traction! However, by relocating workshops and businesses on a human scale, by manufacturing durable goods, by equipping ourselves with a few simple, bust-sized machines, and by preserving a certain number of existing assets – such as digital control machines – we should be able to maintain a good part of current productivity while reducing energy content. These manufacturing units, which are less productive but more labor-intensive and closer to the consumption basins, would be linked to economic networks for the recovery, repair, resale and sharing of everyday objects.

19 1736.

Certainly, in many cases, the objects coming out of such workshops would be more expensive – but not systematically, because the technological enrichment and the multiplication of functionalities also contribute to increasing the cost: a simple, light, frugal and robust low-tech car would certainly cost much less, both to buy and to use[20]. However, the increase in unit costs could be compensated at the level of each household by lower consumption in volume (reuse, repair and energy-saving solutions) and at the societal level by taking better account of negative externalities and an appropriate tax policy (see section 8.4).

8.4. A possible and desirable transition

8.4.1. *The role of the public authority, at all scales*

Such a profound change would require the support and assistance of public authorities at all levels, from the state to the municipalities, via the regions, departments and metropolises.

The state has, first of all, the normative and regulatory leverage. It could easily, for example, enact a weight limit for new vehicles or a ban on many disposable products – it is possible that the recent announcements by the French government on the desire to "break with waste"[21] are along these lines. Although this lever is not always easy to use, as such measures can quickly be considered technocratic, "liberticidal" or aimed at the most fragile populations, it remains an effective tool provided that it is used fairly and exemplarily.

The state and local authorities then have prescriptive power and a very important local support power through deciding on the content of their purchases, orienting the specifications in the direction of sustainability. The initiatives that are currently emerging everywhere could be multiplied, amplified and generalized. Each municipality could open citizen repair centers, "resource-recycling centers", and launch "zero waste" initiatives in administrations, schools and public companies, all of which could be supported by the enthusiasm of a growing part of the population for these themes. Public spending and subsidies could be redirected to promote human

20 See the example of the Dacia Logan model, emblematic of "frugal" innovation.

21 Édouard Philippe, French Prime Minister, "Déclaration de politique générale du 12 juin 2019".

work and the local economy rather than the consumption of energy and raw materials.

Finally, tax choices are structuring. The current system is the result of a mixture of a 19th century paradigm and social consensus dating from 1945. The former considers that resources are "free", or at least worth only the capital and labor needed to extract them. The second has made human labor the foundation of social protection, whether it be healthcare, pensions or unemployment insurance. For example, the French carbon tax represents less than 1% of the social contributions of wages[22], and taxation on the consumption of resources, soil artificialization and the production of hazardous or non-hazardous waste remains very low, if not non-existent.

With such a tax system, the incessant search for productivity becomes "normal" and "natural" for all economic players: administrations are faced with taxpayers who want them to be more efficient and less costly, and companies are faced with their competitors. However, this productivity consists above all in replacing human labor with the consumption of resources and energy.

As available labor is abundant, while resources have become costly from an environmental point of view, it seems necessary to rethink the way to arbitrate between labor and resources, for example by financing social protection through increasingly high environmental taxation. Organizational choices, production methods, project profitability and the usefulness of investments could be profoundly modified, so that they could promote human employment and have beneficial effects on the environment. Employees who are consumers would see their net salary increase, allowing them to pay the additional cost of environmental taxes in the first instance, before gradually shifting to more virtuous and environmentally conscious behavior – provided that the consumption alternatives would be built in parallel.

This would make the objects more repairable than at present, where the price of labor is prohibitive and eco-participation in the purchase ridiculously low. One more employment-intensive alternative, soil- and ecosystem-regenerating agriculture, would become viable and competitive, and its products would be of better quality. Craft trades, small businesses,

---

22 450 billion euros in social security contributions; 50 billion euros in environmental taxes – mainly on fuels – including 4 billion euros for the climate-energy contribution.

renovation and short economic loops, all more intensive in human labor, creators of social ties, rooted in the territories, would find a new lease of life.

8.4.2. *Finding the right scale*

One last point remains to be addressed. The alternative production and consumption system, on the scale of one country or of a few countries wishing to accelerate their transition, should be able to operate in an unchanged international context, including in the intermediate phase – a tax "big bang" being difficult to imagine, a gradual but sufficiently rapid evolution, in the range of 5–10 years, should be considered.

It would indeed, and unfortunately, be illusory to carry out such evolutions on a global and probably European scale, such as calling for a world environment organization that could impose binding measures (e.g. a global carbon price). The slow discussions on climate under the aegis of the UN, or the difficulties of social and fiscal convergence at the European level, have shown that the system of consensus does not allow us to move quickly, nor to take decisions commensurate with what is at stake – not to mention the regular retreats of one or another state, due to internal politics.

To these difficulties must be added legitimate concerns about the deterioration of international competitiveness and the associated risks of job losses (Bihouix 2015). With a carbon tax worthy of the name on a single territory only, for example, energy-intensive industries (steel, cement, chemicals, etc.) would be weakened and subject to stiff competition from imports not subject to the tax.

It would then be necessary to set up customs protections, which are still a real taboo for political elites, who associate protectionism and aggressive nationalism, and economic elites, who have benefited in recent decades from the construction of Europe and globalization. However, reflections are evolving on the subject, and the denunciation, both by certain economic stakeholders and by the associative world, of the latest free trade treaties negotiated by the European Commission (Canada, Mercosur), notably for environmental reasons, shows that the lines are shifting.

The modalities of these customs protections are to be constructed; they could be financial (allowing the integration of an estimated cost of carbon, for example) or normative (considering the social and/or environmental standards of the producing countries). One could also imagine other measures that would encourage local production: kilometer tax, differentiated VAT rates, etc.

8.4.3. *Humans, the key to "repairability" and optimal recycling*

By ceasing to place the "burden" of social protection on human labor, many alternatives that contribute to the preservation of resources could emerge and become viable.

There are less brutal strategies for recycling metals than shredding electronic equipment (or end-of-life vehicles) and plunging them into metallurgical furnaces in even the world's most efficient plants. One can, for example, make the components last by reusing them in refurbished products (refurbishing/remanufacturing) or selling the equipment on the second-hand market, practices that are beginning to develop but require rather skilled labor.

Furthermore, the quality of recycling depends directly on the rate of labor "injected" into the process. The better and easier it is to recover the different metals, the more one will move toward a high level of dismantling, separation and sorting upstream, as this avoids creating mixtures with "metallurgical incompatibilities". Fine disassembly is extremely labor-intensive, although not quite as much as assembly, of course, since there is no need to worry about the esthetic or functional result after the action. However, in our productivist societies, assembly benefits from an extremely high rate of mechanization and robotization, while disassembly, the poor part of the lifecycle, is reserved for waste that goes to countries with low social and environmental standards[23], or is simply substituted by simple shredding after elementary disassembly: schematically, we remove the battery and the screen of telephones, then we shred; we remove the battery, tires and seats of vehicles, then we shred.

A circular economy should be based on a reduction in the volumes of objects manufactured and consumed, possibly resulting in less work upstream of use, while downstream a larger workforce could repair, disassemble, sort and reuse. This will only be possible by making raw materials more expensive.

---

23 The case of electronic waste is emblematic.

Are we willing to pay more for the scarce and poorly recycled resources of our phones and computers? Probably, because a smartphone contains less than €2 of rare metals – half for the gold in the microprocessor, half for the few precious metals such as platinoids. We could pay 10 times more for our resources without too much difficulty! Obviously, the impact would be greater for heavier products. However, more expensive objects would also promote durability, "repairability", modularity, an economy of functionality, and even the transition from car to bicycle!

8.4.4. *Positive impacts*

A low-tech approach could have other, more immediate beneficial effects than the preservation of scarce resources for future generations – a laudable goal, but one for which it is difficult to mobilize populations and governments, as demonstrated by the current inertia on the climate issue.

A transition based on sobriety could ease current and future international tensions over raw materials (cobalt or tantalum in the Democratic Republic of Congo, lithium in the Chile–Bolivia–Argentina triangle, Chinese rare earth elements, platinoids from Russia and South Africa, etc.).

Faced with the risks of dependency that intensify with "green" growth, some people advocate a mining revival in Western countries, with geopolitical, and also economic and environmental arguments – "made in Europe" mining operations would be cleaner than in China, and above all, by having them right before us, we would have to assume the environmental consequences, which would lead us to think and act to reduce our metal waste.

However, the European subsoil is far from containing all the rare metals that our so precious smartphones contain, while a project like the Montagne d'Or, in the Guyanese forest, makes no sense, since barely 10% of the gold mined worldwide is used in industry (mainly electronics), and there is more gold in stock in central banks and private individuals than reserves in the planet's subsoil.

Above all, a transition based on "soft" technologies could contribute to the greater autonomy and resilience of territories, which could prove useful at the dawn of a century that promises to be not that much calmer than the previous ones: less dependence on complex logistics flows, price fluctuations in world markets and less risk linked to future systemic disruptions (constraints in the supply of fossil fuels or rare resources, geopolitics, climate change, etc.). Global dependency would be replaced by increasing local agility.

This would not in any way prevent trade (even though it should, in quantity and value, decrease rather than continue to increase) or "turn in on itself". Would it not rather be to imagine, experiment, build alternative and complementary solutions, introduce diversity, a source of resilience (in the same way that a rich ecosystem resists disturbances better than a poor ecosystem or a monoculture), maintain or rediscover know-how rather than depending solely on captive and uncontrolled – and uncontrollable – technologies because they are too complex?

Which smart city do we want for the day after tomorrow? The one based on digital "intelligence", equipped with sensors manufactured in Southeast Asia, using materials and parts produced in dozens of different (and often unstable) countries, entrusting its data to a Californian multinational that will store them in data centers in Iceland, eminently dependent on a complex world with a structural need for permanent acceleration? Or the one based on the intelligence of its inhabitants, more autonomous and masters of their destiny, in a more stable and peaceful economic system?

8.4.5. *A "happy" transition or nothing*

Human beings have always needed to dream and project themselves into a better future, depending on the case and the time, for themselves or their children, before or after their death. The current techno-utopias are very improbable or scary: pursuit of technological acceleration, "augmented" humanity, galactic conquest, etc. As for the announcements of an inevitable or extremely probable collapse – it is not certain that mobilization is occuring on a scale that would allow a sufficiently radical and rapid change.

If we want to succeed in "landing" our mad ship, an expensive and wasteful world economic system, while preserving some beautiful achievements, such as democracy and fundamental freedoms (at least where they still exist), we will certainly have to find a new utopia for tomorrow and the day after tomorrow; but above all, the changes made will have to create immediately, or as quickly as possible, a better life, a little more happiness, for a maximum number of people. If there is a transition, it will have to be a happy transition.

8.5. References

Bailey, R. (2001). Post-scarcity prophet: Economist Paul Romer on growth, technological change, and unlimited human future [Online]. Available at: https://reason.com/2001/12/01/post-scarcity-prophet-2/ [Accessed March 2021].

Bihouix, P. (2015). Les *low-tech*, emplois de demain. In *Économie de l'après-croissance, Politiques de l'anthropocène II*, Sinai, A. (ed.). Presses de Sciences Po, Paris.

Bihouix, P. (2019). *Le bonheur était pour demain*. Le Seuil, Paris.

Bihouix, P. (2020). *The Age of Low Tech*. Bristol University Press, Bristol.

CGDD (2012). L'Empreinte carbone de la consommation des Français, évolution de 1990 à 2007 [Online]. Available at: https://www.bilans-ges.ademe.fr/static/documents/ressources/fiche%20numero%20124.pdf [Accessed March 2021].

CGDD (2018). L'Empreinte matière, un indicateur révélant notre consommation réelle de matières premières [Online]. Available at: https://www.statistiques.developpement-durable.gouv.fr/lempreinte-matieres-un-indicateur-revelant-notre-consommation-reelle-de-matieres-premieres [Accessed March 2021].

Diamond, J. (2005). *Collapse: How Societies Choose to Fail or Succeed*. Viking Press, New York.

Ehrlich, P. (1968). *The Population Bomb*. MacMillan, London.

Elvis, A. and Milligan, T. (2019). How much of the Solar System should we leave as wilderness? *Acta Astronautica*, 162, 574–580.

Fabrique écologique (2019). Pourquoi et comment développer l'innovation "low-tech" [Online]. Available at: https://www.lafabriqueecologique.fr/vers-des-technologies-sobres-et-resilientes-pourquoi-et-comment-developper-linnovation-low-tech/ [Accessed March 2021].

Fuller, B. (1938). *Nine Chains to the Moon*. Lippincott, Riverwoods.

Kahn, H. and Wiener, A. (1967). *The Year 2000: A Framework for Speculation on the Next Thirty-Three Years*. The Macmillan Company, New York.

Meadows, D.H., Meadows, D.L., Randers, J., Behrens III, W.W. (1972). *The Limits to Growth*. Universe Books, New York.

Nordhaus, W. (1973). The allocation of energy resources. *Brookings Papers on Economic Activity*, 3, 529–575.

Nordhaus, W. (1974). Resources as a constraint on growth. *The American Economic Review*, 64(2), 22–26.

OECD (2018). Global material resources outlook to 2060. Economic drivers and environmental consequences [Online]. Available at: https://www.oecd.org/environment/waste/highlights-global-material-resources-outlook-to-2060.pdf [Accessed March 2021].

Osborn, F. (1948). *Our Plundered Planet.* Little, Brown and Company, New York.

Rifkin, J. (2011). *The Third Industrial Revolution: How Lateral Power is Transforming Energy, the Economy, and the World.* Palgrave MacMillan, New York.

Simon, J. (1981). *The Ultimate Resource.* Princeton University Press, Princeton.

Toffler, A. (1970). *Future Shock.* Random House, New York.

UNEP (2013). *Recycling Metals. Opportunities, Limits, Infrastructure.* UNEP, Nairobi.

UNEP (2019). Global resources outlook 2019. Natural resources for the future we want [Online]. Available at: https://wedocs.unep.org/handle/20.500.11822/27517 [Accessed March 2021].

US Geological Survey (2019). Mineral commodity summaries [Online]. Available at: https://www.usgs.gov/centers/nmic/mineral-commodity-summaries [Accessed March 2021].

Veltz, P. (2017). *La Société hyper-industrielle.* Le Seuil, Paris.

Vogt, W. (1948). *Road to Survival.* William Sloane Associates, New York.

World Bank (2017). The growing role of minerals and metals for a low carbon future [Online]. Available at: https://openknowledge.worldbank.org/handle/10986/28312 [Accessed March 2021].

Conclusion

Florian FIZAINE[1] and Xavier GALIÈGUE[2]
[1] *IREGE, Savoie Mont Blanc University, Annecy, France*
[2] *LEO, University of Orléans, France*

Concluding a book whose ambition is to shed light on complex issues, often presented concisely and carrying considerable stakes, is a difficult task. We hope that by the end of the book, the reader will have had access to a set of information, data and explanatory diagrams that will have made them aware of the issues and challenges that will have to be met if the energy transition is to be compatible with the use of mineral resources. We also hope that in a field where emotional investments and ideological postures are often very marked, the various contributions will have enabled the reader to better understand the complexity of the issues raised by the management of mineral resources, including the energy transition. Finally, we hope to have contributed to a constructive dialog between geosciences, engineering sciences and the humanities in a field where environmental, economic and social issues constantly interact.

Nevertheless, this book has left a number of questions open, a situation that reflects the contingencies of the implementation of a collective work, as well as certain gaps and insufficiencies in French-language research in this field. These gaps and shortcomings, which are the limits of this book, concern each of its parts.

First of all, it is always difficult to give an overall assessment of the mineral resource needs that will result from the energy transition. Such an evaluation would require the use of empirical general equilibrium macroeconomic models like the MIT EPPA6 model, which was the basis of the first Club of Rome report[1]. While the various contributions have provided an initial approach to the situation on all mineral resource markets, the projections that could be made of their demand at the global level would require, in order to be carried out rigorously, the use of one or more macroeconomically self-contained general equilibrium models, which we are still lacking[2]. Reduced forms certainly exist, such as stylized general equilibrium models, but they remain confined to aggregate theoretical models that do not make it possible to obtain numerical forecasts of metal demand according to the different energy transition scenarios. The multi-regional input–output models that have been presented here have taken a first step in this direction, which calls for others. Second, it is equally difficult to arrive at clear-cut results on the environmental and social assessment of mining activity, especially in developing countries. The work presented in this book is undoubtedly a real step forward, but it remains clear that it requires further study, particularly with regard to the long-term social and environmental effects of mineral resource exploitation.

If we want to draw a general message from all the contributions of this book, it would be that, first of all, the energy transition will have a strong material imprint, which requires going beyond the comforting speeches that see in this transition a simple solution to all our problems. Moving from one sociotechnical system to another is not done without delay or resistance, and wanting to free ourselves from the "carbon lock-in" in which our societies remain locked is not simple (Unruh 2000): the gains in well-being acquired, thanks to the free availability of abundant and inexpensive energy sources, are too precious, and behaviors and habits are very difficult to change. The technical and physical constraints related to the disposal of mineral resources also concern their recycling, which can make an undeniable contribution to the reduction of their material footprints but is also confronted with physical and economic limits. In other words, the choice of energy transition is not a single corridor based on a binary logic (more or less decarbonated energy) but can be accomplished, in fact, in different ways, in particular through the use of increased energy efficiency and through a different economic organization of the energy system. Therefore, there could be solutions which

1 See the MIT Joint Program on the Science and Policy of Global Change: https://globalchange.mit.edu/research/research-tools/eppa.

2 Most of the models existing today are based on scenarios where demand is exogenous.

do not systematically rely on more mineral resources. Finally, a global reconfiguration of our way of life could also lead to greater circularity in our consumption and production patterns, which should not prevent us from giving thought to greater frugality in our consumption and production methods. Let us look at each of these points.

C.1. The challenge of access to primary resources

One of the major questions raised by this book is first of all to conjure up what appears more and more as a spectrum, and an illusion, that of the inevitable exhaustion of primary resources. For fossil resources as for mineral resources, we are no longer in the universe of the writings of Jevons or the first reports to the Club of Rome which foresaw a rapid and inevitable exhaustion of fossil and mineral resources on a national and then planetary scale. Prospecting, technical progress and adaptation have reduced this threat, and this depletion has been constantly postponed, but at a time when the much more tangible threat of global warming induced by our greenhouse gas emissions is looming. Now the prospect of not depleting fossil fuels, at least in the short to medium term, has led to a perverse effect, the "Green Paradox", which slows down the transition to renewable energies.

First stated by Sinn (2015), this "Green Paradox" is based on the following observation: knowing that the limits to growth no longer relate to the depletion of fossil fuels, but to the damage that their combustion will cause in the atmosphere, the fight against global warming should require the holders of these resources to *abandon* their extraction in the long term. Far from depleting these resources, they should therefore be voluntarily left in the ground. For these holders of fossil fuels, this means that what is for them an asset of considerable value should no longer have any value, again with a very variable horizon. Therefore, their response will not be to reduce this extraction, but to *increase* it in order to preserve their wealth.

Thus, policies promoting renewable energies can lead to a perverse effect, which is to push fossil fuel producers to *accelerate* their extraction, and thus contribute even more to global warming than if no proactive policy were implemented (for a detailed presentation of the Green Paradox, see the study by Van der Ploeg and Withagen (2015)). Clearly, this Green Paradox is at work, and it allows us to understand the inertia of the global energy system and the persistence of its use of fossil fuels (Smil 2016).

In any case, regardless of the strength of support policies for renewables and the voluntarism of energy policies, fossil fuels will remain present in the global energy mix in 2050, or even in 2100 (Mac Dowell 2015). This resilience of fossil energies could be encouraged by the use of CO_2 capture and storage (CCS) technologies, which is foreseen in the prospective scenarios of international organizations (IEA 2013, 2016), but the large-scale deployment of which is still largely hypothetical (Budinis *et al.* 2018 and Mac Dowell *et al.* 2017).

One consequence of the Green Paradox concerns development policies: faced with the progressive abandonment of fossil fuels, equity between countries with different standards of living should result in the industrialized countries abandoning the rents from these fossil resources to the developing countries, prohibiting their extraction on their territory and privileging it in the developing countries. It must be noted that recent developments are not spontaneously moving in this direction: while some countries, such as France with the Hulot law on hydrocarbons, have moved in this direction, it is often those that are very poorly endowed (and therefore have little to lose by voluntarily abandoning fossil fuel exploration), while those that are abundantly endowed, such as the United States and Canada, have been very careful not to do so, even accelerating this exploitation with polluting and controversial technologies.

For mineral resources, this problem of depletion must be approached from a different angle. We have seen in this book that the availability of mineral resources is much greater than that of fossil fuels, with great disparities according to their relative scarcity, their uses and the possibilities of substitution between products and recycling: even if they may appear for one metal or another, overall shortages will not occur tomorrow, and it will no doubt be disconcerting to compare in half a century's time certain alarmist predictions with their realization. However, the work carried out by Smil (2013) shows that the trend toward an increase in the demand for material goods in our societies has in no way been influenced by the development of a service or digital economy. The transition from concentrated to diffuse energy also requires considerable material investments, whether they concern the equipment necessary for the production of renewable energies or the implementation of new infrastructures. That this evolution is inevitable, and that it will lead to a positive long-term economic[3] and environmental outcome, seems to be a given (Gibon *et al.* 2017), without any doubt, but we have seen that it will not be smooth sailing.

On the other hand, we have also observed that the accessibility of these resources involves considerable economic and geopolitical interests, and that they mobilize an equally considerable amount of energy, with sometimes high regional and local environmental costs. If, once again, the excavating madness and clichés about the plundering of developing countries must be rejected, reflection must also be undertaken on the respect of environmental standards in developing countries, and on the environmental and social responsibility of companies operating in this sector. There remains the problem of artisanal mining, the impact of which was also seen to be much more nuanced than one might think, positive in terms of poverty reduction, much less so in terms of environmental protection. Bringing these artisanal mines out of informality and managing their complementarity with industrial mines is in any case a major challenge for developing countries. Another challenge for developing countries is their insertion in the mineral resource value chain, and their increasing involvement in the transformation of their initial production, some examples of which have been given in various contributions, and which the actors of the mining sector will be led to integrate more and more into their strategies.

C.2. The challenge of circularity and recycling

Another considerable field of research that seems promising is the partial or total recycling of mineral resources. This possibility of recycling differentiates these resources from fossil fuels and thus helps to alleviate the pressure on their primary resources. Here again, the results presented clearly underline the limits of these recycling possibilities, technical limits when the recyclying is of increasingly sophisticated metals and alloys, economic limits when its cost proves too high compared to scrapping or when it concerns materials for which the growth in demand will always require recourse to the primary resource. Subsidy schemes can be used to promote recycling, provided that their cost to the public purse is not too high. Such subsidies are conceivable and desirable when they can support the development of immature technology and encourage behavioral change. In

3 If a wind turbine requires a higher investment cost than a coal-fired power plant, its operating cost will then be much lower.

this way, they allow for economies of scale, learning effects and technical progress that will allow for gradual breakthroughs. Are such trends likely to emerge in the field of recycling, as seems to be the case with policies supporting renewable energies? This is another question that remains open. More often than not, the cost of recycling and its low acceptability in industrialized countries too often lead to the delocalization of these activities in developing countries, with mostly disastrous efficiency and environmental results.

The transition to a circular economy appears in this field as a more global response, making it possible to generalize recycling at all stages of production and use of the material goods at our disposal. As with recycling, it would seem to be illusory to imagine a society operating in a vacuum without any impact over its environment, according to a perpetual movement that defies the laws of thermodynamics. On the other hand, the development of the circular economy would make it possible to strongly reduce the pressure of the economy on its environment, and first and foremost its demand for primary products. However, this can only take place gradually, as the constraints imposed by a model based on abundant and inexpensive energy are so strong, whether they concern the design of the various transport, housing and distribution infrastructures or individual behavior. All things that are often difficult to change are conditioned by the accelerated renewal of products. In this field, technical innovation must be accompanied by social innovation.

This social innovation could be announced as attractive and appealing, especially when it involves a complete reconfiguration of the way we produce and consume. We are thus thinking of solutions that would make it possible to reduce the material footprint of growth, notably through the economy of functionality or the economy of sharing, which unfortunately have not been discussed in this book. We would have liked to welcome contributions from experts on the subject. However, out of the half dozen or so authors solicited, none considered themselves sufficiently available or competent to write on the impact of these new forms of economy in terms of mineral resources. Indeed, although often already advanced as an integral part of the circular economy, the companies and sectors of the economy of sharing or functionality remain marginal and/or in their infancy. In this context, very few impact studies, such as lifecycle analyses (LCAs), have been carried out on the subject and they have often remained confined to energy aspects. This will have to be done, however, because the promises of the circular economy without evidence remain assertions.

C.3. The challenge of reducing our material footprint: decoupling or degrowth?

Contrary to what is often perceived in the media or by the general public, energy transition cannot be reduced to a dynamic of passage from one energy extraction system to another. If its ultimate objective is indeed the fight against global warming and the reduction of the use of fossil fuels, the energy transition must go beyond the mere decarbonation of energy to mobilize the three other levers for reducing greenhouse gas emissions that can be seen through the now famous Kaya identity, namely:

– population decline;

– increasing energy efficiency;

– degrowth.

Let us consider the first point mobilized by the Kaya identity, that of the population. While it seems certain that the demographic transition will come to an end with a leveling off and then a decrease in world population by the end of the century, this will not be sufficient to compensate for the growth of the second element of the Kaya identity, GDP per capita, the growth of which will remain high, particularly in emerging countries – China and India now, Africa tomorrow. This has important implications for the consumption of mineral resources and is what leads authors such as Smil (2013) to reject the hypothesis of a perpetuation of the material growth rates observed to date. If the population lever, because of its inertia, cannot be mobilized in the short to medium term, in the long term the developments observed in industrialized countries allow us to envisage a decrease in the population which would thus make it possible to alleviate the anthropic pressure on natural resources. This also raises the problem of the aging of societies which will experience a reverse demographic transition, a situation already experienced in some countries such as Japan, Germany and Bulgaria, which are from this point of view laboratories for this unprecedented social fact.

If we move on to the second lever of energy efficiency, for a long time a large number of economists have adopted the reassuring vision presented by Kuznets and his famous curve. This inverted U-curve, initially applied to income inequalities, holds that the spontaneous functioning of a liberal economy initially leads to an increase in inequalities, and then to their reduction beyond a certain turning point. It is therefore not surprising that environmental and material versions of these inverted U-curves have appeared through the initial work of Malenbaum (1978) on metals, Grossman and Krueger (1991, 1995) on the environment, and De Bruyn and Opschoor (1997) and Bleischwitz *et al.* (2018) on the material footprint of growth. These studies suggest that growth, after a certain threshold, will spontaneously become "greener" or "cleaner". It is not our purpose here to discuss the details of the arguments evoked in relation to Kuznets curves. Quite simply, in the field of inequality, recent developments seem to largely invalidate their predictions, as the work of Piketty and Milanovic has amply demonstrated. In the field of environment and natural resources, too, the most successful examples of current work and indicators reject the existence of a natural emergence of solutions as we become richer (Wiedmann *et al.* 2015; Pothen and Welsh 2019).

In the area of energy transition, a majority of public policies have fought against global warming by mobilizing the decarbonation of energy through the use of renewable energy and, to a lesser extent, energy efficiency. However, we know little about the material footprint of energy efficiency operations (no exhaustive LCA-type study has been published to date). However, the kWh or TJ of electrical or thermal energy avoided could potentially prove to be more lifesaving than its decarbonation. However, the gains from improving energy efficiency and decarbonation are very often offset by the rebound effect highlighted several times in this book.

Already evoked by the premonitory work of W.S. Jevons, the rebound effect is omnipresent in all areas of environmental economics, and contributes to the invalidation of the Kuznets curve. How many innovations to improve energy performance in housing or in transportation have enabled us to improve our comfort, increase our travel and discover new activities? It is clear that the only technical response to environmental challenges proves to be in vain, simply because it is confronted with humanity's infinite desire for wealth, to paraphrase the title of Daniel Cohen's book (2015), and the frustrations of the most deprived. This rebound effect can also concern the circular economy, as we have seen in this book, or the economy of sharing, because it can help to satisfy new desires at a lower cost. Does not carpooling already allow some people to envisage new journeys that would not have taken place in its absence? Fighting the rebound effect requires breaking the causality between the increase in the productivity of our resources and that of our consumption, in other words, achieving an absolute decoupling between our enrichment and our environmental footprint, in all its dimensions. This is only possible if we are able and determined to individually change our behavior in favor of greater sobriety, which is something that can be done first by those whose desires are already satisfied!

For the others, those whose needs are not being met, whether in developed or developing countries, this message is harder to hear, as they suffer frugality in increasingly unequal societies. Another option, more collective, aims to give a strong signal in terms of price or quotas on resources, energy and the environment (where the price of carbon is only one illustration among others). Yet again, our unequal and short-sighted societies stumble over the cost borne by the weakest and over measures considered to be liberticidal: why and how can we impose something that others will pay for later? The limits of technology and its inability to respond alone to moral and ethical choices are thus clearly apparent here.

If sobriety is so appealing on paper but rarely put into practice, it is perhaps because there are also reasons that go far beyond these relatively technical motifs. For example, even if our government would voluntarily orient its policy toward moderation, the isolated application of a policy of "imposed" sobriety through the imposition of standards or taxes would probably be overwhelmed by pro-growth economic policies in neighboring countries. France is not an isolated nation in space, and it is highly likely that people will go elsewhere (abroad in this case) to seek statutory goods that give them a sense of elevation and domination over their fellow citizens. This desire for positioning vis-à-vis others seems to be a legacy of the evolution of our brains and would be one of the five factors implicitly guiding our actions via the striatum and bypassing our cortex (Bohler 2019).

Indeed, within this analytical framework our unconscious actions would appear to be biased by food, gender, effort reduction, social status and information seeking. However, the occasional and continuous satisfaction of most of these factors progresses partially with economic growth. We would then be stuck in an analogy of Nash's prisoner's dilemma, where all players know that the natural expression of their choice will lead them to a bad solution (the pursuit of growth and endless desires) and also that the isolated action of a stakeholder toward a good solution (renunciation and sobriety) will not result in the realization of this choice at the general level but rather in its relative downgrading with respect to others. Guaranteeing this sobriety through self-isolation by protectionism also seems illusory. We do not know of any country that has succeeded in closing its borders durably while maintaining its economic and social gains. Once again, however, no defeatism is necessarily essential because the contribution of the human sciences could upset the situation by showing how this state of affairs could be changed by education and culture. It would be a matter of making desirable and promoting a way of life that requires fewer economic services and material and energy resources. Other scientists say no different when they speak of breaking the relationship between well-being and economic services rendered, and thus moving from a measure of standard of living, itself broader than a measure of wealth, to a measure of quality of life (Von Weizsäcker *et al.* 2013). These are open reflections that will have to find answers if we want the energy and ecological transitions of our economies to fully achieve their objective and move away from a purely technical response.

This is also a major challenge that the energy transition will have to face, the most difficult one, because it goes beyond technical solutions to raise major economic, political and social issues. For some, the current failure of decoupling therefore requires us to dare to consider scenarios of economic (and material) decline, which few economists have done so far. If the thirst for growth is as strong as ever, how can we convince the poorest to abandon their hopes and adopt a lifestyle that is more respectful of their well-being? Economists still continue to look where there is light and data, whereas we should take a step aside and study the relationships between resource consumption and other indicators of capabilities, according to Sen's (1999) concept. Some premises for this work do exist (Steinberger and Roberts 2010; Steinberger *et al.* 2012; Mayer *et al.* 2017), but we need to go further in order to sketch out future sustainable paths that are possible from a resource point of view, sustainable for the environment and desirable for humans. This is a task that researchers from all disciplines will no doubt have to tackle as quickly as possible, and one that will not fail to provide us with a few more fine discoveries.

C.4. References

Bleischwitz, R., Nechifor, V., Winning, M., Huang, B., Geng, Y. (2018). Extrapolation or saturation – Revisiting growth patterns, development stages and decoupling. *Global Environmental Change*, 48, 86–96.

Bohler, S. (2019). *Le Bug humain, pourquoi notre cerveau nous pousse à détruire la planète et comment l'en empêcher*. Robert Laffont, Paris.

Budinis, S., Krevor, S., Mac Dowell, N., Brandon, N., Hawkes, A. (2018). An assessment of CCS costs, barriers and potential. *Energy Strategy Reviews*, 22, 61–81.

Cohen, D. (2015). *Le monde est clos et l'univers infini*. Albin Michel, Paris.

De Bruyn, S.M. and Opschoor, J.B. (1997). Developments in the throughput-income relationship: Theoretical and empirical observations. *Ecological Economics*, 20, 255–268.

Gibon, T., Arvesen, A., Hertwich, E.G. (2017). Life cycle assessment demonstrates environmental co-benefits and trade-offs of low-carbon electricity supply options. *Renewable and Sustainable Energy Reviews*, 76, 1283–1290.

Grossman, G.M. and Krueger, A.B. (1991). Environmental impacts of a North American free trade agreement [Online]. Available at: https://doi.org/10.3386/w3914 [Accessed 25 May 2021].

Grossman, G.M. and Krueger, A.B. (1995). Economic growth and the environment, 353–377 [Online]. Available at: https://doi.org/10.2307/2118443.

IEA (2013). *Technology Roadmap: Carbon Capture and Storage*. IEA, Paris.

IEA (2016). *Twenty Years of Carbon Capture Storage*. OECD/IEA, Paris.

Mac Dowell, N., Fennell, P., Shah, N., Maitland, G.C. (2017). The role of CO_2 capture and utilization in mitigating climate change. *Nature Climate Change*, 7, 243–249 [Online]. Available at: https://doi.org/10.1038/nclimate3231.

Malenbaum, W. (1978). *World Demand for Raw Materials in 1985 and 2000*. McGraw-Hill, New York.

Mayer, A., Haas, W., Wiedenhofer, D. (2017). How countries' resource use history matters for human well-being – An investigation of global patterns in cumulative material flows from 1950 to 2010. *Ecological Economics*, 134, 1–10.

McGlade, C. and Ekins, P. (2015). The geographical distribution of fossil fuels unused when limiting global warming to 2°C. *Nature*, 517, 187–190 [Online]. Available at: https://doi.org/10.1038/nature14016.

Pothen, F. and Welsch, H. (2019). Economic development and material use. Evidence from international panel data. *World Development*, 115, 107–119.

Sen, A.K. (1999). *Development as Freedom*. Anchor Books, New York.

Sinn, H.W. (2008). Public policies against global warming: A supply-side approach. *International Tax and Public Finance*, 15, 360–394.

Sinn, H.W. (2015). Introductory comment – The Green Paradox: A supply-side view of the climate problem. *Review of Environmental Economics and Policy*, 9(2).

Smil, V. (2013). *Making the Modern World. Materials and Dematerialization*. John Wiley & Sons, Hoboken.

Smil, V. (2016). Examining energy transitions: A dozen insights based on performance. *Energy Research & Social Science*, 22, 194–197.

Steinberger, J.K. and Roberts, J.T. (2010). From constraint to sufficiency: The decoupling of energy and carbon from human needs, 1975–2005. *Ecological Economics*, 425–433.

Steinberger, J.K., Roberts, J.T., Glen, P.P., Baiocchi, G. (2012). Pathways of human development and carbon emissions embodied in trade. *Nature Climate Change*, 2, 81–85.

Unruh, G.C. (2000). Understanding carbon lock-in. *Energy Policy*, 28, 817–830.

Van der Ploeg, F. and Withagen, C. (2015). Global warming and the green paradox: A review of adverse effects of climate policies. *Review of Environmental Economics and Policy*, 9(2), 285–303.

Von Weizsäcker, E., Hargroves, K., Smith, M., Desha, C., Stasinopoulos, P. (2013). *Facteur 5 : comment transformer l'économie en rendant les ressources 5 fois plus productives*. De Boeck, Louvain-la-Neuve.

Wiedmann, T.O., Schandl, H., Lenzen, M., Moran, D., Suh, S., Wrest, J., Kanemoto, K. (2015). The material footprint of nations. *Proc. Natl. Acad. Sci. USA*, 112, 6271.

List of Authors

Charlène BARNET
IFP Énergies nouvelles (IFPEN)
Rueil-Malmaison
France

Philippe BIHOUIX
AREP
Paris
France

Samuel CARCANAGUE
Institut de relations internationales et stratégiques (IRIS)
Paris
France

Michel DESHAIES
LOTERR
University of Lorraine
Nancy
France

Florian FIZAINE
IREGE
Savoie Mont Blanc University
Annecy
France

Xavier GALIÈGUE
LEO
University of Orléans
France

Alain GELDRON
Agence de l'environnement et de la maîtrise de l'énergie (ADEME)
Angers
France

Fernanda GUEDES
IFP Énergies nouvelles (IFPEN)
Rueil-Malmaison
France

Emmanuel HACHE
IFP Énergies nouvelles (IFPEN)
Rueil-Malmaison
and
Institut de relations internationales et stratégiques (IRIS)
Paris
France

Thierry LEFÈVRE
Center for Research on Advanced Materials (CERMA)
Laval University
Quebec
Canada

Stephanie REICHE-DE VIGAN
Paris Sciences et Lettres University
France

Gondia SOKHNA SECK
IFP Énergies nouvelles (IFPEN)
Rueil-Malmaison
France

Johan YANS
Institut Transitions
Lyon
France
and
ILEE
University of Namur
Belgium

Index

3Rs (Reduce, Reuse, Recycle), 164, 173

A, B

alloy, 129, 167, 175, 177–179, 197
Antarctic, 40, 57, 63–65
Arctic, 64
bauxite, 14, 30, 111, 145
boundaries, 50, 59, 61, 92, 122, 125, 129, 139, 140, 154, 157, 158, 171, 177, 189
  planetary, 139, 140, 157
by-product, 7

C

cartelization, 11, 13, 14, 16, 17
Chicago Mercantile Exchange, 25
chrome, 129, 179, 194
citizen participation, 117
climate, 18, 31, 76, 80, 139, 194, 201, 202
  change, 66, 67, 165, 188, 205
Club of Rome, 4, 173, 188
co-production, 132
cobalt, 12, 13, 15, 18, 19, 25, 30, 31, 115, 171, 191, 193, 194, 204
code minier, 44, 47
collapse, 4, 188, 189, 205
concrete, 12, 133, 143, 170, 179
conflicts, 4, 51, 56, 59, 89, 90, 92, 93, 95, 114, 122, 164
constraints, 48, 76, 121, 122, 129–131, 133–135, 163, 167, 171, 176, 179, 190, 194, 205
consumption, 11, 13, 16, 20, 26, 77, 92, 121, 126, 128, 139–141, 145–147, 149, 150, 152–158, 167, 172, 181, 187, 190, 191, 194, 195, 197, 199–202
convention, 56, 60–65, 156
  Montego Bay, 60–63, 65
  United Nations, 56, 59, 62, 64
  Wellington, 63–65
copper, 12, 13, 15, 18–21, 25, 27, 28, 30, 31, 47, 73, 74, 77, 79, 80, 81, 87–90, 121, 122, 124, 127–129, 131, 134, 143, 170, 171, 175, 178, 179, 188, 190, 191, 193, 195, 198

cornucopians, 121
criticality, 5–8, 10–13, 26

D

decoupling, 139, 141, 142, 149, 150, 154, 155, 157, 158, 190–192
dematerialization, 154, 191
density of the population, 85
dependence/dependency, 4, 5, 7, 8, 11, 12, 19–22, 27, 31, 91, 108, 131, 152, 165, 193, 204
  path, 131
domestic material consumption (DMC), 145
doomsdayers, 187, 188
downcycling, 167, 195

E

eco-design, 189, 195, 198
economy, 5, 6, 11, 17, 31, 48, 56, 78, 82, 85, 90, 94, 107, 108, 114, 118, 121, 122, 128, 141, 142, 147, 149, 152, 154, 157, 158, 163, 164, 166, 170, 173, 181, 182, 189, 190, 194–196, 201, 203, 204
  circular, 48, 107, 157, 170, 173, 189, 190, 194, 195, 203
  local, 94, 201
  of functionality, 204
ecosystem, 157, 181, 189, 201
efficiency, 25, 122, 133, 134, 147, 150, 154, 157, 168, 173–175, 177–179, 194, 196, 197
  energy, 3–6, 11–14, 25, 26, 30, 31, 57, 122, 154, 157
employment, 83, 108, 176, 182, 201, 202
energy, 26, 28, 48, 49, 54, 77, 78, 81, 82, 122, 126–128, 131, 135, 142, 154, 158, 176, 178, 181, 190–194, 197–199, 201, 202
environment, 6, 27, 40, 42, 44, 48, 49, 55, 57, 64–66, 71, 74, 78, 90, 92–94, 118, 122, 133, 140, 175, 182, 188, 189, 196, 201, 202
environmental regulation, 50
exploitation, 9, 16, 39, 40, 42, 44, 45, 47–49, 51–53, 56–67, 71–76, 78–82, 85, 87–95, 103, 106, 112, 114, 118, 140, 149, 165, 167, 195, 196
  quotas, 17
exploration, 16, 40, 42, 44, 45, 47, 49, 52, 56, 61, 63–67, 93, 106
exportation, 9, 14, 21
extraction, 18, 19, 22, 23, 26, 28, 49, 72, 76–78, 80–82, 85, 86, 89, 90, 103–106, 108–113, 115–118, 128, 146, 165, 171, 176, 181, 188, 191–193, 196
extraterritorial spaces, 40

F

footprint, 140, 141, 146, 153–155, 157, 192
  ecological, 140, 154
  environmental, 141
  material, 133, 134, 140, 141, 146, 153, 155, 157, 146, 149, 152
free trade, 21, 202
frugality, 195

G

gallium, 106, 115, 195
GDP, 31, 123, 140, 141, 146, 147, 149–154, 158, 191, 192
geographical redistribution, 81
geological, 5–7, 11–13, 44, 72, 105, 109, 115–117
geopolitical, 3–8, 10–13, 27, 30, 31, 114, 118, 157, 204, 205
germanium, 106, 195
Global Political Risk Index, 7
global warming, 5, 49, 86, 149
gold (metal), 47, 51, 71, 73, 77, 79, 89–92, 104, 107, 111, 124, 127, 131, 133, 154, 164, 167, 179, 195, 204
growth, 3, 13, 17, 20, 23, 51, 73, 75–77, 79, 81, 90, 103, 118, 140, 141, 147, 149, 150, 152–155, 157, 158, 170, 171, 173, 187, 189–193, 196, 204
  demographic, 3, 103, 147, 187
  economic, 51, 79, 141, 147, 149, 153, 154, 157, 188, 190
  green, 157

H, I

Herfindahl Hirschman Index, 7
impact, 23, 25, 66, 71, 72, 85, 92–94, 108, 113, 121, 131, 133–135, 145–147, 149, 150, 152, 154–156, 175, 188, 194, 196, 204
indium, 106, 115, 175, 193, 195
intensity, 135, 146, 149, 187
international area, 40, 56, 57, 59, 61, 64, 66
investment, 6, 31, 55, 117, 123–125, 130, 166, 176, 179
iridium, 132
iron, 19, 21, 73, 77, 79, 81, 89, 104, 105, 127, 135, 143, 145, 164, 166, 171, 190, 191, 193

J, K, L

joint production, 6
Kuznets curve, 153
law, 26, 39, 48, 56, 173
  gas, 39
  mining, 39, 40, 42, 46
lead, 19, 28, 47, 73–76, 79, 104, 106, 107, 129, 132, 133, 175, 182, 188, 191, 193–195
legal, 39, 45, 48, 50, 56–58, 60, 63, 65
lifecycle analysis, 6, 135
lithium, 12, 13, 15–18, 25, 30, 31, 81, 189, 193, 204
local skills, 109
London Metal Exchange, 25
low-tech, 187, 195, 200, 204

M

markets, 4–6, 11, 13, 14, 16–18, 23, 25, 30, 31, 108, 131, 133, 135, 163, 176, 188, 205
material flow accounting (MFA), 142
materials
  mineral, 5, 42, 47, 103, 104, 106
  raw, 4–6, 8, 11, 13, 14, 16–21, 24–31, 105–107, 109, 123, 142, 145–147, 163, 164, 171, 176, 182, 188, 190, 192, 201, 203, 204

mineral
  industrial, 152, 163
  non-metallic, 140, 143
  resources, 21, 30, 39, 40, 42–52, 54, 56–58, 61–67, 77, 81, 91, 103, 108, 115, 116, 122, 127, 128, 134, 141
mining, 18, 19, 21, 44, 45, 74, 80, 83, 84, 86–91, 93–95, 105, 111–115, 117, 154
modurality, 204
monetarization, 181
moratorium, 40

N, O, P

nickel, 12, 18, 19, 28, 31, 124, 134, 191, 193
NIMBY syndrome, 109, 110, 115, 118
nuisance, 92, 93, 103, 118, 181
ore, 10, 26, 76, 80, 88–90, 105, 107, 111, 112, 115, 117, 165, 166, 170
palladium, 124, 129, 131, 132, 179, 195
platinum, 123, 124, 129, 132, 133, 195
polymetallic nodules, 62, 63
population, 3, 18, 27, 31, 71–74, 77, 79, 80, 84, 85, 95, 84, 112, 140, 142, 146, 152, 154, 156, 196, 200
price elasticity, 123
primary, 132
property, 42–47, 55, 59, 165
  mineral, 42
protection, 30, 40, 42, 44, 48–50, 52, 54, 55, 61, 64, 94, 182, 201, 203

R

R/P ratio, 6
rare earths, 12, 15, 18, 22–25, 28, 31, 81, 127, 131, 132, 134, 173, 176, 179, 195, 204
rate, 7, 13, 83, 84, 128, 140, 141, 144, 145, 157, 167, 168, 170–174, 178, 189, 190, 193, 196, 198, 203
  efficiency rate of recycling process, 168
  end-of-life waste collection, 168
  end-of-life waste recyling, 168
  metabolic, 145, 147, 152
  waste incorporation into production, 168
rebound effect, 154, 157, 197
recycling, 3, 7, 26, 67, 103, 121, 144, 157, 163–168, 170–179, 181, 182, 191, 194, 195, 198, 203
rehabilitation, 94, 112, 197
relocation, 25, 49
repairability, 203, 204
reserve, 42–44, 65, 115
resource curse, 91
responsibility, 64, 67, 93, 94, 142, 154–156, 158, 164, 165, 173
  extended producer, 164, 165
reuse, 26, 133, 163, 176, 200

S

saturation (of demand), 152–154
sea, 40, 56–62, 64
seabed, 40, 56–58, 61–66
short circuit, 107, 108, 118
silver, 47, 71, 73, 74, 76, 77, 79, 89, 90, 107, 130, 132, 135, 188, 195

sobriety, 25, 187, 195, 196, 204
social
  acceptability, 67, 111
  progress index, 158
socioeconomic metabolism, 142, 143
sovereignty, 52–54, 56, 58–61
stock, 128, 142, 189, 195, 196, 204
strategic, 3, 12, 22, 28, 115
substitution, 109, 121–135
supply, 7–9, 16–18, 20–24, 26–28, 31, 49, 57, 75, 77, 94, 105, 106, 108, 109, 118, 131, 157, 171, 173, 205
sustainability, 66, 122, 126, 127, 188
sustainable management, 39, 44, 50, 51, 65–67

T

tantalum, 179, 195, 204
technologies, 4, 5, 16, 28, 30, 31, 50, 114, 125, 127, 131, 133, 163, 172, 193, 195, 196, 204, 205
  low-carbon, 4, 5, 31
tellurium, 193, 195
territory, 3, 17, 19, 20, 25, 27, 31, 46, 48, 54, 55, 64, 77, 104, 106, 170, 182, 192, 202
tin, 19, 27, 47, 73, 79, 127, 133, 175, 178, 194
trade, 20, 52, 54, 77, 145, 147, 153, 192
  international, 20, 52–56, 58, 62, 77, 145, 147, 153
transition, 3–6, 11–14, 25, 26, 30, 31, 60, 103, 114, 122, 126, 147, 152, 195, 200, 202, 204, 205
treaty, 40, 49, 54, 64
tungsten, 107, 109

V, W, Z

volatility, 7, 13, 16, 17, 29
vulnerability, 7, 22, 28
Word Governance Index, 7–9
zinc, 9, 19, 28, 47, 104–107, 129, 178, 188, 193–195